Particle Technology and Surface Phenomena in Minerals and Petroleum

Particle Technology and Surface Phenomena in Minerals and Petroleum

Edited by

Mahendra K. Sharma
Eastman Chemical Company
Kingsport, Tennessee

and

G. D. Sharma
Rogaland University Centre
Stavanger, Norway
and University of Alaska, Fairbanks
Fairbanks, Alaska

Springer Science+Business Media, LLC

Library of Congress Cataloging in Publication Data

Particle technology and surface phenomena in minerals and petroleum / edited by Mahendra K. Sharma and G. D. Sharma.
p. cm.
"Proceedings of The Fine Particle Society Symposium . . . held August 21–25, 1991, in San Diego, California"—T. p. verso.
Includes bibliographical references and index.
ISBN 978-1-4899-0619-9
1. Flotation—Congresses. 2. Oil fields—Production methods—Congresses. 3. Surface chemistry—Congresses. 4. Particles—Congresses. I. Sharma, Mahendra K. II. Sharma, G. D. (Ghanshyam Datt), 1931– . III. Fine Particle Society.
TN523.P27 1992 91-39706
622'.752—dc20 CIP

Proceedings of The Fine Particle Society Symposium
on Particle Technology and Surface Phenomena in Minerals and Petroleum,
held August 21–25, 1991, in San Diego, California

ISBN 978-1-4899-0619-9 ISBN 978-1-4899-0617-5 (eBook)
DOI 10.1007/978-1-4899-0617-5

PREFACE

The current state of the art of several aspects of minerals and petroleum is presented in this volume. It documents the proceedings of the Internationl Symposium on Particle Technology and Surface Phenomena in Minerals and Petroleum sponsored by the Fine Particle Society (FPS). This meeting was held in San Diego, California, August 21-25, 1990. The symposium upon which this volume is based was organized in five sessions emphasizing various basic and applied aspects of research on minerals and petroleum technology. Major topics discussed involve surface phenomena in minerals, mineral flotation, characterization of aspaltenes, theoretical aspects of reservoir simulation, porosity, permeability, residual water saturation, hydrocarbon and gas potential in north slope Alaska, polymer/alkaline flooding, surfactant flooding and foam flooding for enhanced oil recovery.

This edition includes eighteen selected papers presented in the symposium. These papers are divided in four broad categories: (1) Asphaltene Aggregation and Characterization, (2) Theoretical Aspects and Reservoir Characterization, (3) Colloidal Dispersions in Minerals/Petroleum, and (4) Surface Phenomena and Petroleum Recovery. Several mineral and oil recovery processes using various chemicals with special reference to surface phenomena and particle technology are described in these sections.

This proceedings volume includes discussions of various processes occuring at molecular, microscopic, and macroscopic levels in mineral flotation and petroleum recovery processes. The editors hope that this volume will serve its intended objective of reflecting the current understanding of formulation and process problems related to minerals and petroleum recovery processes. In addition, it will be a valuable reference source for both novices as well as experts in the field of minerals and petroleum technology. It will also help the readers to understand underlying surface phenomena and will enhance the reader's potential for solving critical formulation and process problems.

The editors would like to convey their sincere thanks and appreciation to the Fine Particle Society for the generous support that allowed them to invite many researchers from several countries to participate in the symposium. We would also like to express our thanks and appreciation to Ms. Patricia M. Vann and to the Editorial Staff of the Plenum Publishing Corporation for their continued interest in this project.

The editor are grateful to reviewers for their time and efforts in providing valuable comments and suggestions to improve the material presented in the manuscripts. We wish to convey our sincere thanks and appreciation to all authors and coauthors for their contributions, enthusiasm and patience. The views and conclusions expressed herein are those of the authors.

One of us (MKS) would like to express his thanks to the appropriate management of the Eastman Chemical Company (ECC) for allowing him to participate in the organization of the symposium and to edit this proceedings volume. His special thanks are due to Mr. J. C. Martin (ECC) for his cooperation and understanding during the tenure of editing this proceedings volume.

Finally, MKS wishes to express his sincere thanks to his colleagues and friends for their assistance and encouragement throughout this project. Also he would like to acknowledge the assistance and cooperation of his wife, Rama, and extends his appreciation to his children (Amol and Anuj) for allowing him to spend many evenings and weekends working on this volume.

M. K. Sharma
Research Laboratories
Eastman Chemical Company
Kingsport, TN 37660

G. D. Sharma
Petroleum Development Lab.
University of Alaska
Fairbanks, AK 99775
or
Rogaland University Center
Stavanger, Norway

CONTENTS

ASPHALTENE AGGREGATION AND CHARACTERIZATION

THEORETICAL ASPECTS AND RESERVOIR CHARACTERIZATION

THE ROLE OF ASPHALTENE AGGREGATION IN VISCOSITY VARIATION OF RESERVOIR HYDROCARBONS AND IN MISCIBLE PROCESSES

V. A. Kamath, M. R. Islam, S. L. Patil,
J. C. Jiang and M. G. Kakade

Petroleum Development Laboratory
University of Alaska Fairbanks
Fairbanks, Alaska

The variation of oil viscosity with depth and/or location has been reported in many reservoirs around the world. This paper examines the role of asphaltene aggregation in the variation of viscosity of reservoir fluids. We conclude that the viscosity of heavy oils is dependent upon the extent of asphaltene aggregation rather than asphaltene concentration alone. A modified Einstein equation has been used to predict the oil viscosity as a function of asphaltene concentration and molecular weight which govern the extent of asphaltene aggregation.

The role of asphaltene deposition in miscible flooding processes is also examined. Experimental data together with coupled equation of state models and Flory-Huggins polymer solution theory have been used to illustrate the effect of various parameters such as solvent type, solvent/oil ratio and pressure on the amount of asphaltene precipitation during addition of solvents to heavy oil.

COMPOSITIONAL DEPENDENCE OF VISCOSITY OF BITUMENS AND HEAVY OILS

In many petroleum reservoirs around the world, reservoir fluid composition has been found to very with location and depth.[1-3] Patel[4] found the viscosity of Athabasca, Peace River, Wabasca and Cold Lake bitumens to vary with depth of the formation. Schulte[3] explained the compositional variations within a hydrocarbon column by gravity segregation phenomenon. However, he found that the extent of variation to be higher with larger aromatic fractions in the hydrocarbon fluid. Hirschberg[5] concluded that the heavy polar components play a key role in compositional and oil viscosity variation and in particular, identified asphaltene segregation to have a dominant effect. Hirschberg found that the

Particle Technology and Surface Phenomena in Minerals and Petroleum
Edited by M.K. Sharma and G.D. Sharma, Plenum Press, New York, 1991

viscosity of the North African reservoir oil sample increased by factor of 4 when the asphaltene content increased form 10% to 16%.

The viscosity data for Mexican crude[6], Peace River bitumen[7,8], Athabasca bitumen[9], California crude[10] all show that the viscosity of oil and asphalt blends increase with the concentration of asphaltenes. Altgelt and Harle[11] also studied the effect of asphaltene on asphalt viscosity. They found asphaltenes to form aggregates in solution. The degree of which was found to depend upon structure, molecular weight and concentration of the asphaltenes and the solvent power. They concluded that the viscosity of asphaltene is primarily due to this aggregation.

Numerous studies on the effect of temperature on bitumen viscosity have been conducted by various researchers in recent years and a number of viscosity correlations have been proposed. However, due to the complexities involved in determining the extent of asphaltene aggregation, not much attention has been paid to the compositional dependence of viscosity of bitumens and heavy oils.

The classical Einstein's equation[12] relates the viscosity of an infinitely dilute suspension of solid spherical particles to the viscosity of the dispersion medium as follows:

$$\frac{\mu}{\mu_o} = 1 + K_E\phi \tag{1}$$

where μ is the viscosity of the suspension, μ_o is the viscosity of dispersion medium, ϕ is the volume factor of the solids and K_E is Einstein coefficient. This equation however does not consider the molecular weight as a separate variable. Staudinger[13] modified the Einstein equation to incorporate the effect of molecular weight of solute particles on the viscosity ratio. To incorporate the solute particle interactions, Gillespie[14] suggested the replacement of ϕ by ϕ_{eff} defined as follows:

$$\phi_{eff} = a\phi + Ka^2\phi^2 \tag{2}$$

where a is a function of number of particles in aggregate, aggregate shape and the packing factor of the particles in the aggregates and K is the parameter which is equal to 0.9 for a suspension of unidisperse and unisize solid particles. It increases with an increase in the axial ratio of aggregates and decreases if the particles are of different size. In this study the Einstein equation was modified as follows:

$$\left(\mu/\mu_o\right) - 1 = CMWa\phi\left(1 + Ka\phi\right) \tag{3}$$

where MN is the average molecular weight of asphaltenes, ϕ is the volume fraction of asphaltenes precipitated, α, K and C are constants which need to be determined by fitting experimental viscosity data.

Experimental viscosity data of Altgelt and Harle[11] was used to fit the model and to obtain the values of α, K and C for asphaltenes of different molecular weights which are presented in Table 1. The coefficient C is considered to be a function of the maltene fluid. For this data, since the maltene fluid was the same, the C is constant. the coefficients α and K varied with asphaltene molecular weight. Figure 1 shows experimental data of Altgelt and Harle[11] and the predictions of Equation 3. As can be seen from this figure, the predictions match the experimental data very well. Thus, such an approach may be useful for the prediction of viscosity variations in reservoirs. Further experimental data are required to obtain generalized parameters. The volume fractions of asphaltene aggregates in this equation can be determined from the static asphaltene precipitation tests of from the coupled equation of state model and the Flory-Huggins polymer solution theory as described later.

Table 1. Coefficients C, α and K obtained by Fitting Experimental Viscosity Data of Altgelt and Harle[11]
C = 0.0006968

Molecular Weight	α	K
960	10.00	89
8000	6.76	50
20000	7.85	50
25000	10.0	49.9

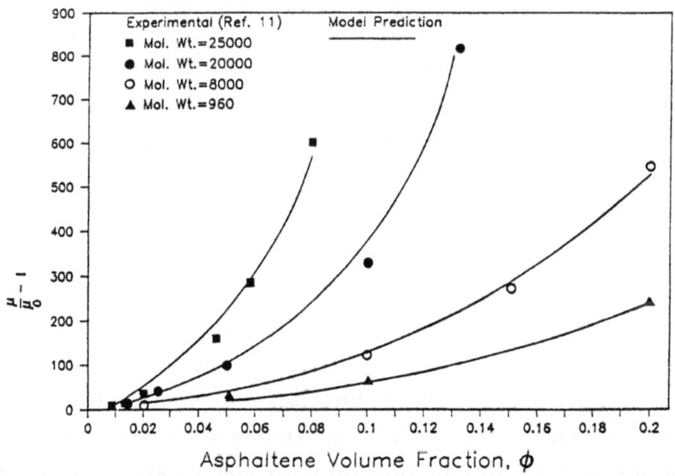

Figure 1. Comparison of Experimental Viscosity Data with Model Predictions

ASPHALTENE PRECIPITATION DURING MISCIBLE DISPLACEMENT PROCESSES

It is well known that flocculation of asphaltenes in petroleum reservoirs, wells and surface separation-upgrading facilities pose technical problems and increase the cost of production and processing of crudes. Field conditions conducive to precipitation of asphaltenes include: natural depletion, miscible flooding, caustic flooding, acid stimulation and gas-lift operations.[15-18] Asphaltene precipitation is particularly important problem in miscible flooding since it can reduce permeability, affect well injectivities and productivities, alter rock wettability characteristics and even cause plugging of producing wells.[19,20]

In miscible processes, enriched hydrocarbon gases, LPG's or CO_2 (termed as solvents) are injected into the reservoir to reduce oil viscosity and interfacial tension, to provide drive mechanism and to achieve (dynamic or direct contact) miscibility with the crude thus resulting into very high displacement efficiency and ultimate oil recovery. In such processes however, the miscible solvents can also cause precipitation of asphaltenes by altering the resin to asphaltene ratio of the crude. The amount of asphaltene precipitation depends upon the crude composition (the original asphaltene and resin content of the crude), the solvent composition, the pressure and temperature conditions and the phase behavior of solvent-oil mixtures. Asphaltene deposition in the reservoir is even more complex phenomenon since several processes such as solvent mixing, fingering, mass transfer between oil and solvent in a multi-contact process, adsorption of asphaltenes on preferential rock sites occur simultaneously. Detailed investigations into asphaltene precipitation have been impeded by a shortage of experimental data needed to quantify asphaltene precipitation.

West Sak reservoir located on the North Slope of Alaska is estimated to contain up to 25 billion barrels of heavy oil in place and represents the largest known heavy oil accumulation in the United States. The possibility of sharing the existing Kuparuk River Unit facilities makes the development and production of the West Sak reservoir, a near-term target. The absence of natural drive mechanism in this reservoir makes it a target for the application of enhanced oil recovery processes. Miscible flooding is considered as one of the candidates for recovery of West Sak crude.[21]

In this study, experimental data on the amount of asphalt and asphaltene precipitation due to addition of solvents to West Sak crude were gathered. The first set of tests were conducted for West Sak stock tank oil. Solvents used include: ethane, carbon dioxide, propane, n-butane, n-pentane, n-heptane, Prudhoe Bay natural gas (PBG) and natural gas liquids (NGL). Effect of solvent to oil dilution ratio on the amount of precipitation was studied. Alteration of crude oil composition due to asphalt precipitation was measured using gas-liquid chromatography. Second set of experiments were conducted to measure asphaltene precipitation due to addition of CO_2 to live (recombined) West Sak crude.

Also a thermodynamic model based on the coupled Equation of State model and Flory-Huggins theory for polymer solutions was developed. The model parameters such as solubility-parameter of asphaltenes, molecular weight of asphaltenes, and molar volume of asphaltenes were obtained by fitting the model to experimental data.

Asphaltene Precipitation from Tank Oils

Asphaltene/asphalt precipitation tests were conducted for West Sak (tank) oil using eight different solvents. Composition of West Sak (tank) oil is given in Table 2. Solvents used include: carbon dioxide, ethane, propane, n-butane, n-pentane, n-heptane, Prudhoe Bay natural gas (PBG) and natural gas liquids (NGL). The compositions of PBG and NGL are given in Table 3.

Following procedure was used to measure asphalt and asphaltene precipitation. Initially, a known amount of tank oil was taken in a visual cell. In case of liquid solvents such as n-pentane and n-heptane, the known amount of solvent was directly added to the tank oil in the cell. In case of volatile solvents such as ethane, propane, n-butane, CO_2 and NGL, the solvents were first condensed to liquid state and then known amount of solvent was added to the tank oil in the cell. After preparation of a solvent-oil mixture in the cell, the cell contents were continuously agitated with a rocking mechanism for several hours at room temperature. The cell contents were then flashed and remaining liquid was filtered to separate precipitated solids. The solids were then thoroughly dried and weighed. In case of n-heptane, the solids precipitated represent asphaltenes. For all other solvents, the solids precipitated represent asphalts. Additional n-heptane was then added to the asphalts to redissolve resins leaving behind only asphaltenes. The asphaltenes after thorough drying were then weighed. These experiments were repeated for each solvent at varying solvent/oil dilution ratio.

Effect of Solvent/Oil Ratio

The amount of asphaltenes precipitated from various solvents are summarized in Table 4. Figures 2-9 show that, in general, increase in solvent/oil ratio increases the amount of asphaltene precipitation and levels off after certain solvent/oil ratio. Addition of solvent to oil reduces concentration of resins (which act as peptizing agents for asphaltenes) in the oil resulting in asphaltene

Table 2. Compositional Changes (Mol %) and Density Changes in West Sak Crude (Tank Oil) after Asphaltene Precipitation by N-Pentane and N-Butane

Component	Original Dead Oil	After Prec. with N-Butane	After Prec. with N-Pentane	Component	Original Dead Oil	After Prec. with N-Butane	After Prec. with N-Pentane
$C - C_6$	-	-	-	C_{23}	1.87	1.73	2.13
C_7	1.27	1.19	0.16	C_{24}	1.22	1.67	1.39
C_8	1.05	2.01	0.28	C_{25}	1.75	2.06	0.96
C_9	1.35	2.03	0.75	C_{26}	1.60	1.84	1.58
C_{10}	1.47	2.11	1.41	C_{27}	1.25	1.46	1.52
C_{11}	1.94	2.74	1.94	C_{28}	1.23	1.59	0.72
C_{12}	1.97	3.27	2.85	C_{29}	1.55	1.21	0.91
C_{13}	2.49	3.45	3.28	C_{30}	0.69	1.17	0.90
C_{14}	2.67	3.49	3.53	C_{31}	2.11	1.38	1.10
C_{15}	2.30	3.20	3.42	C_{32}	2.11	3.02	2.55
C_{16}	2.35	3.29	3.64	C_{33}	2.52	2.70	2.33
C_{17}	2.80	3.74	4.28	C_{34}	1.45	3.22	2.12
C_{18}	2.55	3.34	3.89	C_{35}	2.63	1.14	1.90
C_{19}	2.44	3.10	3.54	C_{36}	4.14	3.62	2.72
C_{20}	1.90	2.38	2.68	$C_{37}+$	40.92	27.55	35.86
C_{21}	2.43	2.55	2.79		100.00	100.00	100.00
C_{22}	1.98	2.75	2.87	Sp. Gravity	0.9390	0.9125	0.9242

Table 3. Compositional Analysis (Mol %) of Live (Recombined) Oil, Prudhoe Bay Natural Gas (PBG) and Natural Gas Liquids (NGL)

Component	Live Oil	PBG	NGL
N_2	0.03	trace	0.02
CO_2	0.02	12.73	trace
C_1	38.33	72.80	trace
C_2	0.86	7.44	trace
C_3	0.36	3.94	1.50
C_4	0.18	2.29	50.50
C_5	0.06	0.80	29.13
C_6	0.20		18.85 (C_{6+})
C_7	0.02		
C_8	0.01		
C_9	0.82		
C_{10}	1.50		
C_{11}	1.72		
C_{12}	1.35		
C_{13}	1.50		
C_{14}	1.80		
C_{15}	1.94		
C_{16}	1.80		
C_{17}	1.57		
C_{18}	1.80		
C_{19}	2.47		
C_{20}	2.84		
C_{21} +	38.82		
Mol. Wt. of Plus Fraction	455		96

Table 4. Comparison of Experimental and Predicted Amounts of Asphaltene Precipitation for West Sak Tank Oil

Solvent Used	Solvent/Oil Ratio by Weight	V_L (ft^3/lb mole)	δ_L $(psia)^{0.5}$	M_L (lb/lbmole)	Wt % Asphaltenes (experimental)	Wt % Asphaltenes (predicted)	% Error
Ethane	3.0	6.20	196.41	363.4	5.14	5.14	0.00
	13.0	6.24	196.54	366.6	5.28	5.28	0.00
CO_2	3.60	6.27	196.55	368.4	5.56	5.50	1.09
	13.60	6.30	196.62	370.8	5.80	6.03	3.97
	15.30	6.30	196.65	371.2	5.89	5.72	2.88
Propane	4.70	5.84	196.05	340.5	8.87	8.34	6.03
	7.80	5.85	196.07	341.2	8.60	9.05	5.02
	16.30	5.87	196.13	342.9	9.07	9.14	0.82
n-Butane	2.00	4.46	193.98	250.9	6.22	6.20	0.34
	8.13	4.47	194.03	251.4	6.36	6.43	1.09
	13.30	4.47	194.08	252.0	6.45	6.40	0.76
n-Pentane	2.00	2.19	182.44	98.56	1.79	-	-
	7.00	1.92	178.27	80.20	6.20	6.47	4.39
	12.50	1.87	177.28	76.71	6.30	5.37	14.78
	20.20	1.85	179.70	74.99	6.69	7.38	10.37
n-Heptane	2.00	2.85	185.27	132.3	3.83	-	-
	7.30	2.53	182.08	130.8	5.34	5.41	1.29
	13.50	2.47	181.32	106.8	7.50	7.05	6.06
	20.10	2.45	181.03	103.8	7.32	7.67	4.81
NGL	3.20	2.33	183.07	105.3	5.73	-	-
	7.60	2.12	179.87	91.1	8.80	8.80	0.00
	9.90	2.08	179.23	88.4	8.88	8.88	0.00
Prudhoe Bay Gas	1.59	6.19	196.41	362.7	5.49	6.22	13.32
	1.70	6.19	196.43	372.8	7.40	6.24	15.66
	5.85	6.22	196.49	364.6	8.15	7.54	7.54
	10.42	6.24	196.55	366.2	8.08	8.48	5.01

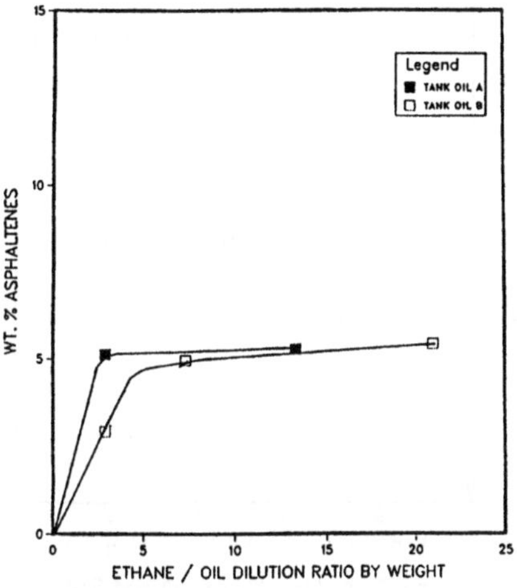

Figure 2. Ethane - West Sak Oil

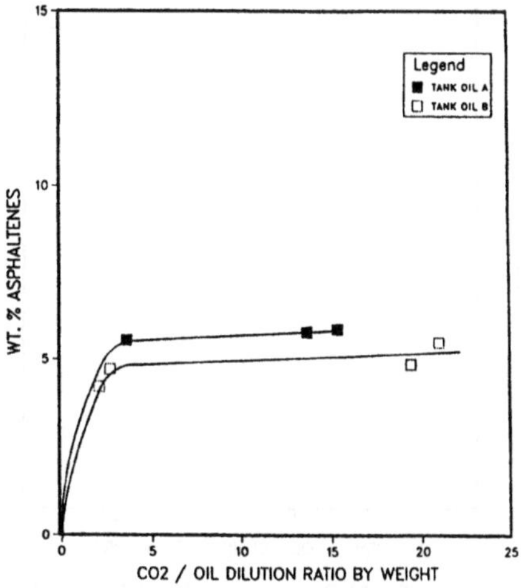

Figure 3. CO_2 - West Sak Oil

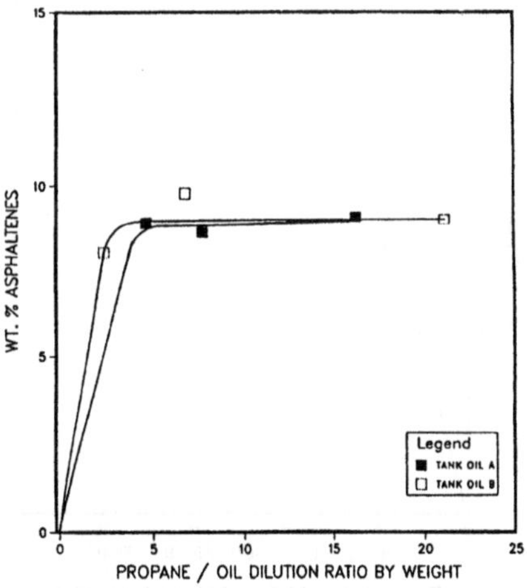

Figure 4. Propane - West Sak Oil

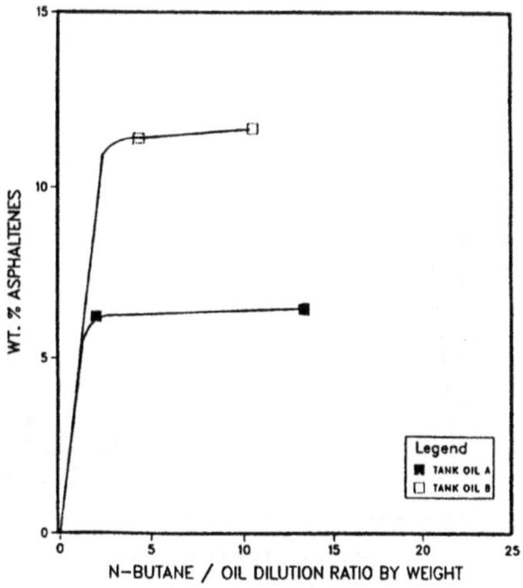

Figure 5. N-Butane - West Sak Oil

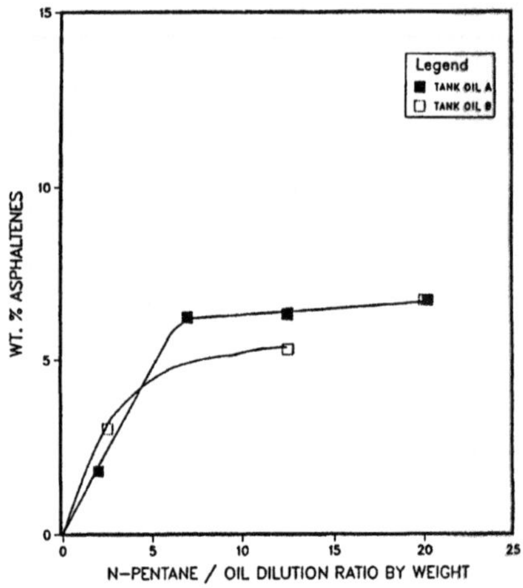

Figure 6. N-Pentane - West Sak Oil

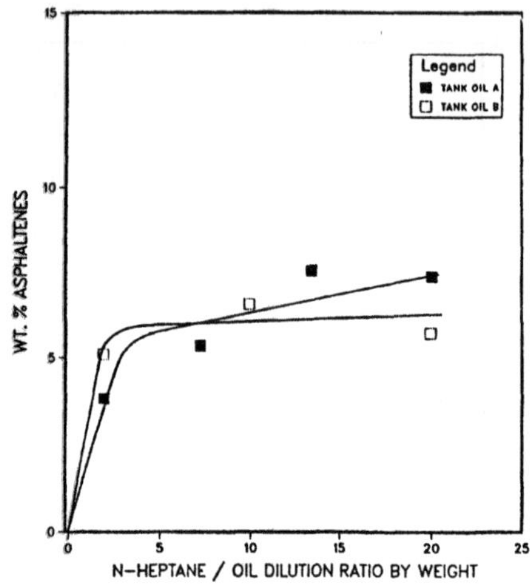

Figure 7. N-Heptane - West Sak Oil

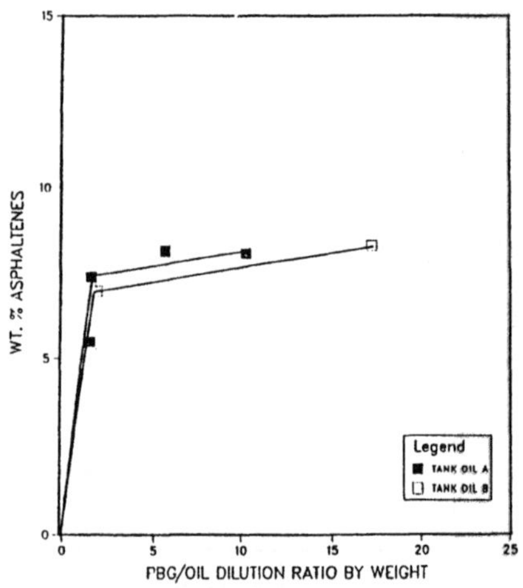

Figure 8. Prudhoe Bay Gas - West Sak Oil

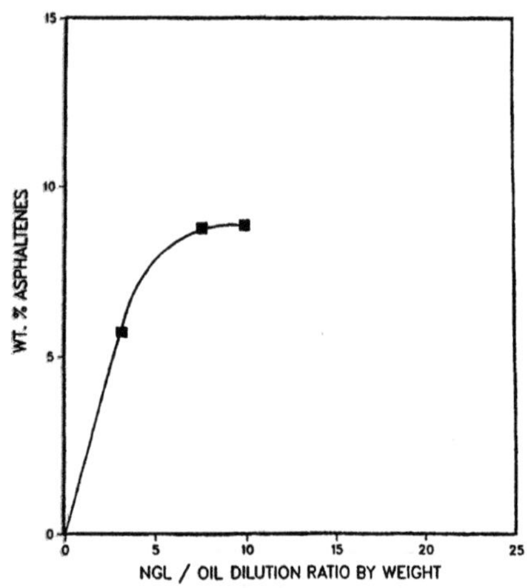

Figure 9. NGL - West Sak Oil

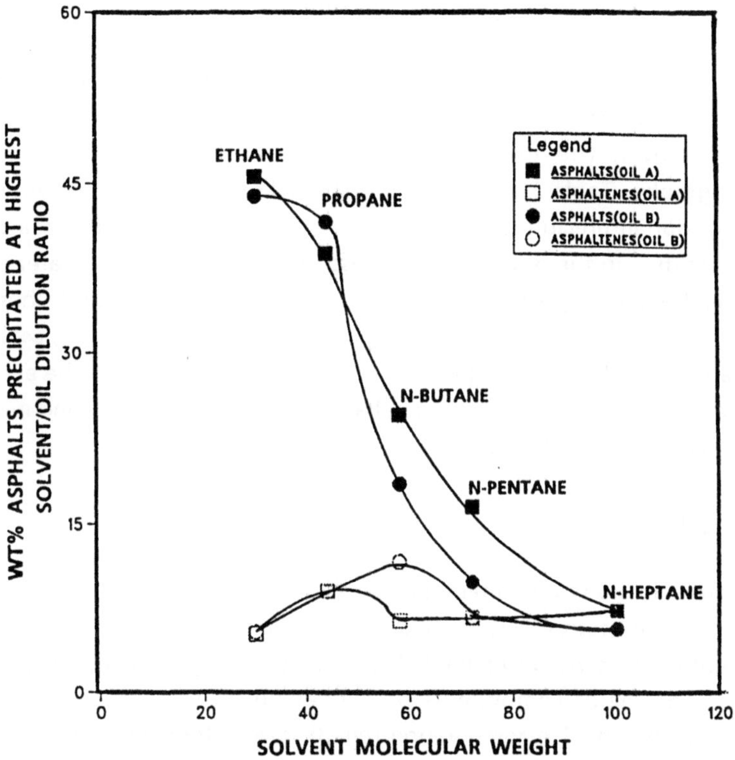

Figure 10. Effect of Solvent Molecular Weight on Amount of Asphalt and Asphaltene Precipitation

precipitation. Figure 10 shows that for normal paraffinic solvents, the amount of asphalts precipitated (at maximum solvent/oil dilution ratio) increase with decrease in molecular weight of solvents while the amount of asphaltenes precipitated remain relatively same. The mutual solubility of resins (which are part of asphalts) in oil decreases with decrease in molecular weight of solvent added. The variation of amount of asphaltene precipitated for each solvent is probably due to differences in the range of asphaltenes that were precipitated by each solvent.

Alteration of Crude Oil Composition Due to Precipitation

In two of the precipitation tests, the asphalts were first precipitated by addition of n-butane and n-pentane to the tank oil. After separation of asphalts from the crude, the deasphalted crude was analyzed by gas-liquid chromatography. Table 2 shows that the original oil composition is considerably altered by asphalt precipitation. The amount of heavier ends in the deasphalted crude is less than in the original crude. Also, n-butane removed greater amount of heavier fraction ($C_{37}+$) than n-pentane. The deasphalted crude has lower density than original crude. This indicates that in a miscible process, asphaltene precipitation can alter the composition of crude oil which needs to be accounted for in prediction of solvent-oil phase behavior, compositional path and miscibility conditions for solvent-oil systems.

Asphaltene Precipitation from Live Oil

In order to study asphaltene precipitation from live oil systems at reservoir conditions and to study effect of pressure on amount of asphaltene precipitation, experimental set up was assembled. Figure 11 is a schematic diagram of the set-up used. The major components of the set up include: 1) JEFRI positive displacement pump, 2) recombination cell, 3) solvent transfer (pistoned) cell, 4) stainless steel (60 micron) in-line filter, 5) air bath equipped with temperature controller and rocking mechanism. Positive displacement pump was used either to inject gas during recombination of oil or to inject CO_2 at constant pressure into recombination cell to prepare desired CO_2 - recombined oil mixture.

Initially, the known amount of tank oil was taken into the recombination cell. The dead oil was recombined at 1705 psia and 80°F. The composition of live oil is given in Table 3. The solvent (CO_2) was then injected from the solvent transfer cell into the recombination cell (which also acted as asphaltene cell) at constant pressure. The amount of CO_2 injected was determined from the pump reading. The resulting CO_2-oil mixture was rocked for several hours to achieve good mixing. The mixture was then passed through a high pressure stainless steel

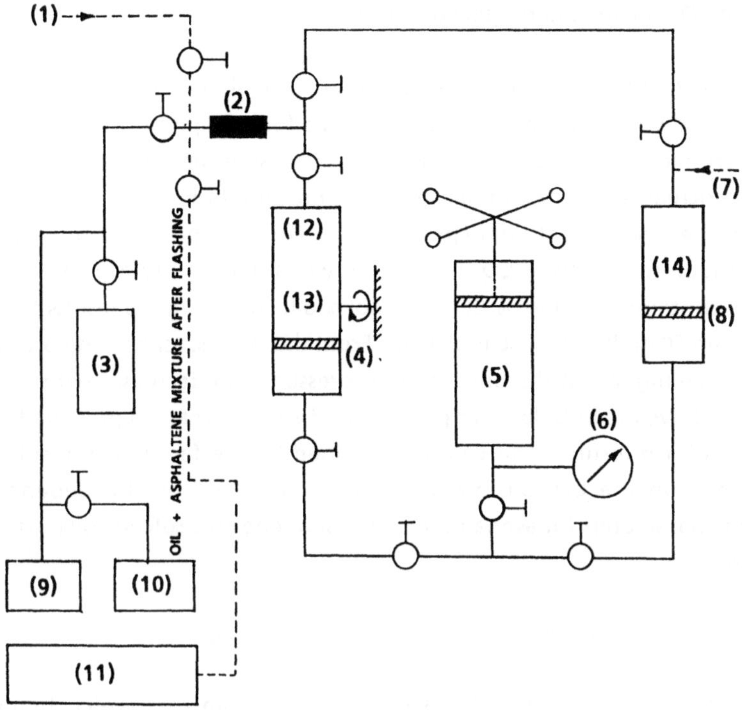

(1)	DEAD OIL VESSEL	(8)	SOLVENT VESSEL
(2)	INLINE FILTER	(9)	GAS CHROMATOGRAPH
(3)	FLASHED GAS VESSEL	(10)	GASOMETER
(4)	RECOMBINATION CELL	(11)	LIQUID CHROMATOGRAPH
(5)	DISPLACEMENT PUMP	(12)	GAS
(6)	PRESSURE GAUGE	(13)	CRUDE OIL
(7)	GAS CYLINDER	(14)	SOLVENT

Figure 11. Schematic of Experimental Apparatus for Asphaltene Precipitation Tests

in-line filter to trap precipitated solids allowing filterate to pass through. The deasphalted crude was then flashed into a oil-gas separator. The amount of asphaltenes in the filter and oil were then measured using some procedure as described earlier.

Effect of CO_2/Oil Ratio and Effect of Pressure

In the first set of experiments, amount of asphaltenes precipitated by adding CO_2 to oil in various proportions at 1705 psia (bubble point pressure of oil) and 80°F were measured. Table 5 shows that increase in amount of CO_2 in CO_2-Oil mixture increases the amount of asphaltene precipitation.

In the second set of experiments, effect of pressure on asphaltene precipitation from a fixed CO_2-oil mixture was studied. Table 6 gives results for mixture containing 30 mol% CO_2 - 70 mol% oil. Table 7 gives results for mixture containing 50 mol% CO_2 - 50 mol% oil. From the results, asphaltene precipitation seems to be highest at the bubble point pressure and decreases with increase in pressure above the bubble point pressure and with decrease in pressure below the bubble point pressure. This can be explained by the fact that the solubility of asphaltenes in the CO_2-oil mixture will be governed by the dependance of solubility parameters for asphaltene and asphaltene-free solvent-oil mixtures with pressure.

Theoretical Basis of Asphaltene Precipitation Modelling

Asphaltenes in crude oil are large, bulky molecules which behave like polymer molecules. Thus, liquid-liquid polymer solution theories are used to

Table 5. Effect of CO_2 Addition to Recombined West Sak Oil on the Asphaltene Precipitation (at 1705 psia, 80°F)

MOL% CO_2 IN CO_2/OIL MIXTURE	WT% ASPHALTENES
0	3.14
20	3.23
40	3.99
60	4.79
80	5.89

Table 6. Effect of Pressure on Amount of Asphaltene Precipitation
30 Mol% CO_2 - 70 Mol% West Sak Crude
(Recombined at 1705 psia and 80°F)

PRESSURE (PSIA)	WT% ASPHALTENES IN THE IN-LINE FILTER	WT% ASPHALTENES IN THE OIL
900	0.177	4.472
1300	0.183	4.631
1705	0.197	4.833
2500	0.151	4.569
3000	0.179	4.963
3500	0.183	3.817
4000	0.212	3.389

Table 7. Effect of Pressure on Amount of Asphaltene Precipitation
50 Mol% CO_2 - 50 Mol% West Sak Crude
(Recombined at 1705 psia and 80°F)

PRESSURE (PSIA)	WT% ASPHALTENES IN THE IN-LINE FILTER	WT% ASPHALTENES IN THE OIL
1000	0.140	4.170
1705	0.192	6.430
3000	0.156	4.750
4000	0.108	4.510

describe phase behavior of asphaltenes, where the remaining components of the crude can be considered as solvent phase in which asphaltenes are either dissolved or suspended. Inherent assumption in such modeling is that the process of asphaltene precipitation is reversible and asphaltenes behave like a non-associating liquid. To model the effect of pressure, temperature, oil composition, and solvent composition on the amount of precipitation, it is necessary to couple the liquid/liquid polymer solution theory to the vapor-liquid equilibria of oil-solvent mixtures.

Two approaches were considered to describe asphaltene phase behavior. The first approach considers asphaltenes as a single homogeneous compound represented by an average molecular weight and other asphaltene properties. Such an approach has been used by Hirshberg et al.[22] and Burke et al.[23] and is used in this study as well. The second approach considers that asphaltene particles exist in wide ranges of sizes and molecular weights and treats asphaltenes as a poly dispersed heterogeneous polymer. In such modelling it is necessary to treat asphaltenes by a continuous distribution function.[17] While the second approach is more generalized, it requires a large amount of extensive experimental data for fitting model parameters.

LIQUID-LIQUID POLYMER SOLUTION THEORY

In order to apply Flory-Huggins polymer solution theory to asphaltene-crude oil solutions, it is necessary to treat the crude as a mixture of two liquid phases, the first phase is pure asphaltene liquid phase which acts as a solute and the second phase as the remaining components of crude oil which acts as a solvent phase.

The volume fraction ϕ_A of dissolved asphaltenes is given by following expression:

$$\Phi_A = exp\left[\frac{V_A}{V_L}\left(1 - \frac{V_L}{V_A} - \frac{V_L}{RT}(\delta_A - \delta_L)^2\right)\right] \quad (4)$$

where V_A = molar volume of asphaltenes
V_L = molar volume of liquid phase
δ_A = solubility parameter of asphaltenes
δ_L = solubility parameter of liquid phase.

The solubility parameter is defined as a measure of cohesive energy density or the internal pressure that is exerted by the molecules within a solution. It is obtained from the following definition:

$$\delta_i = \left(\frac{\Delta U_i^V}{V_i}\right)^{1/2} \quad (5)$$

where ΔU_i^v is the internal energy function and V_i is the molar volume of component i. The calculation of solubility parameters is discussed later.

The value of the solubility parameter is characteristic of a specific molecule. When two liquids possessing considerably different solubility parameters are mixed together, the internal pressure exerted by the liquid with higher solubility parameters basically squeeze the molecules of lower solubility parameter liquid out of solution matrix, resulting in the immiscibility. When the solubility parameters of two liquids approach each other, the mutual solubility also increases.

VAPOR-LIQUID EQUILIBRIA CALCULATIONS

In order to determine the solubility parameter (δ_L) and molar volume (V_L) for solvent-oil mixtures at a given pressure and temperature conditions, vapor-liquid equilibrium calculations are performed using Peng-Robinson Equation of State.

Prior to using the Peng-Robinson Equation of State for vapor-liquid equilibria calculations, the Equation of State parameters for each component were tuned using PVT data for oil. Once the Equation of State has been tuned, the mixture of crude oil and solvent is flashed at desired pressure and temperature to determine the composition of liquid phase. Coat's PVT simulator was used for this purpose.[24]

A separate Peng-Robinson Equation of State function program was written to determine the parameters δ_L and V_L given by Equations 6 - 9.

$$V_L = \frac{Z_L RT}{P} \tag{6}$$

$$\frac{da_L}{dT} = \sum_i \sum_j x_i x_j (1 - \delta_{ij})(a_i a_j T_{c_i} T_{c_j})^{1/2} \left\{ \left[1 + K_j \left(1 - \sqrt{T_{r_j}} \right) \right] \left[-\frac{K_i}{2} \sqrt{T_{r_i}} \right] + \left[1 + K_i \left(1 - \sqrt{T_{r_i}} \right) \right] \left[-\frac{K_j}{2} \sqrt{T_{r_j}} \right] \right\} \tag{7}$$

Then

$$\Delta U_L^V = \left(\frac{a_L}{2\sqrt{2}\, b_L} - \frac{T}{2\sqrt{2}\, b_L} \frac{da_L}{dT} \right) ln \left(\frac{V_L + 2.414 b_L}{V_L - 0.414\, b_L} \right) \tag{8}$$

The solubility parameter δ_L is then obtained by

$$\delta_L = \left(\frac{\Delta U_L^V}{V_L} \right)^{1/2} \tag{9}$$

Model Performance

The model consists of three parameters namely, (V_A) the molar volume of asphaltenes, (δ_A) the solubility parameter of asphaltenes and (M_A), the molecular weight of asphaltenes. With the knowledge of these three parameters it is possible to predict the amount of precipitation of asphaltenes that will occur due to addition of a given amount of solvent to crude oil at any pressure and temperature conditions.

To test the model performance, experimental data gathered with Tank Oil were used to fit these 3 parameters. An optimization program was written to obtain the three parameters. For the initial guesses of the parameters V_A and δ_A literature values were used.[22,23] Method of steepest ascent was used to minimize the following objective function, F:

$$F = \frac{1}{N_{DATA}} \sqrt{\sum_{i=1}^{N_{DATA}} \left(1 - \frac{(wt\%\,Asp)_{pred}}{(wt\%\,Asp)_{exp}}\right)^2} \tag{10}$$

where N_{DATA} is no. of experimental data points, $(wt\%\,Asp)_{exp}$ is experimental value of weight % asphaltene precipitated and $(wt\%\,Asp)_{pred}$ is predicted value.

Table 4 shows the comparison between experimental and predicted amounts of asphaltene precipitation for various solvents such as ethane, CO_2, propane, n-butane, n-pentane, n-heptane, NGL and PBG respectively after optimization of the three parameters.

The values of optimized parameters for asphaltenes are given in Table 8. The model predictions in general are in close agreement with the experimental values of asphaltene precipitation. However, there are three problems with this model, i.e. Equation 4. First, it was not possible to obtain one set of parameters for all solvent-oil mixtures. This is expected since each solvent may precipitate a different range of asphaltene fractions of the crude oil. Hence optimization was done to fit parameters for each solvent separately. The second problem is that the model performance is poor for those solvent-oil systems which have considerable variation in V_L and δ_L parameters due to addition of solvent to oil. This is due to the exponential nature of Equation 4. For example, in Table 4 (i.e. for solvents n-pentane, n-heptane and NGL) the addition of solvent changes the V_L and δ_L values for liquid phase, thus low solvent/oil ratio's were not used in regression of asphaltene parameters. But for other solvents such as ethane to n-butane, CO_2, and PBG, the addition of solvents did not change V_L and δ_L considerably and thus all solvent/oil dilution ratios could be used in regression of parameters. The third problem observed with the model is that the parameters fitted with low pressure experimental data are not useful in predicting asphaltene precipitation at higher pressures. Again, this is due to alteration of V_L and δ_L with pressure. Thus, model improvement is needed to account for the effect of pressure on the asphaltene precipitation.

Table 8. Optimized Values of Asphaltene Solubility Parameter, Asphaltene Molar Volume and Asphaltene Molecular Weight For Various Solvents (West Sak Tank Oil)

SOLVENT	δ_A $(psia)^{0.5}$	V_A (ft^3/lbmole)	M_A (lb/lbmole)
Ethane	219.69	14.48	2491
Carbon Dioxide	219.71	14.48	1531
Propane	220.57	14.48	3080
N-Butane	208.06	5.17	1100
N-Pentane	216.84	3.76	1605
N-Heptane	219.89	6.68	2483
NGL	217.00	4.21	1468
PBG	202.70	6.37	1738

CONCLUSIONS

1. Asphaltene precipitation results in removal of heavier fractions of the crude leaving behind lighter deasphalted crude. This indicates that in a miscible process, the alteration of crude composition needs to be accounted for in prediction of phase behavior, compositional path and miscibility conditions for solvent-oil systems.
2. Amount of asphalt precipitation increases with decrease in molecular weight of normal paraffinic solvents.
3. Increase in solvent/oil dilution ratio increases the amount of asphaltene precipitation until a certain dilution ratio after which it levels off.
4. A thermodynamic model based on Flory-Huggins polymer-solution theory was developed and coupled with Equation of State model to predict the amount of asphaltene precipitation. The model prediction shows close agreement with the experimental data after regression of asphaltene properties such as molar volume, solubility parameter and molecular weight. The model, however, fails to account for the effect of large changes in the solubility parameters of the oil-solvent mixtures.
5. The aggregation and segregation of asphaltene primarily influence the viscosity of heavy oils and bitumens and as such can be used to explain variations of viscosity of reservoir hydrocarbons.

ACKNOWLEDGEMENT

The financial support for this work was provided by the U.S. Department of Energy, the Petroleum Development Laboratory, and the U.S. Bureau of Mines, and is gratefully acknowledged. The fluid samples used in this study were provided by CONOCO, Inc. and the help is acknowledged.

REFERENCES

1. Sage, B. H., Lacey, W. N., "*Gravitational Concentration Gradients in Static Columns of Hydrocarbon Fluids*", Trans. AIME, 132, 12-131, 1939.
2. Hunt, M., "*Petroleum Geochemistry and Geology,*" W. H. Freeman and Co., San Francisco, pp. 281-291, 1979.
3. Schulte, A. M., "*Compositional Variations Within a Hydrocarbon Column Due to Gravity,*" SPE 9235, 1980.
4. Patel, M. A., "*Determination of Viscosities of Oil from Mannville Formation Oil Sand,*" Alberta Research Council Report, October, 1973.
5. Hirschberg, A., "*The Role of Asphaltenes in Compositional Grading of a Reservoir's Fluid Column,*" SPE 13171, 1984.
6. Mack, C., "*Colloid Chemistry of Asphalts,*" J. Phys. Chem., 36, 2901, 1932.
7. Waxman, M. H., Deeds, C. T., Closmann, P. J., "*Thermal Alteration of Asphaltenes in Peace River Tars,*" SPE 9510, 1980.
8. Kitzan, P., Parson, L. J., Quoted in "*The Thermodynamic and Transport Properties of Bitumens and Heavy Oils,*" AOSTRA, 1984.
9. Dealy, J. M., "*Rheological Properties of Oil Sands Bitumens,*" Can. J. of Chem. Eng., 57, 677-683, 1979.
10. Data, P, "*Rheological Studies on Asphaltic Crude Oils of California,*" M.Sc. Thesis, University of California, 1962.
11. Algelt, K. H., Harle, O. L., "*The Effect of Asphaltenes on Alphalt Viscosity,*" Ind. Eng. Chem. Prod. Res. Dev., 14(4), 240-246, 1975.
12. Einstein, A., Ann. Phys. Leipzig 19, 289, 1906.
13. Ber., 63, 222, 1930; quoted by Mack, C., J. Phys. Chem., 36, 2901, 1932.
14. Gillespie, T. "*The Effect of Aggregation and Particle Size Distribution on the Viscosity of Newtonian suspensions,*" J. Colloid and Interface Science, 94(1), 1983.
15. Haskett, C. E. and Tartera, M., "*A Practical Solution to the Problem of Asphaltene Deposits - Hassi Messahoud Field, Algeria,*" J. Pet. Tech., pp. 387-391, April, 1965.
16. Lichaa, P. M., "*Asphaltene Deposition Problem in Venezuela Crudes - Usage of Asphaltene in Emulsion Stability Oil Sands,*" SPEJ, pp. 609-624, 1977.
17. Kawanaka, S., Park, S. J., and Mansoori, G. A., "*The Role of Asphaltene Deposition in EOR Gas Flooding: A Predictive Technique,*" SPE/DOE Paper #17376, SPE/DOE Enhanced Oil Recovery Symposium held in Tulsa, OK, pp. 617-625, April 17-20, 1988.
18. Harvey, M. T. Jr., Shelton, L. J., and Kelm, C. H., "*Field Injectivity Experiences with Miscible Recovery Projects Using Alternate Rich Gas and Water Injection,*" J. Pet. Tech., pp. 1051-1055, Sept., 1977.
19. Danesh, A., Krinis, D., Henderson, G. D. and Peden, J. M., "*Asphaltene Deposition in Miscible Gas Flooding of Oil Reservoirs,*" Ind. Chem. Eng. Prod. Res. Dev., Vol. 66, No. 7, pp. 339-344, 1988.
20. Stalkup, F. I., "*Miscible Displacement,*" SPE Monograph, June, 1983.
21. Sharma, A. K., Patil, S. L., Kamath, V. A. and Sharma, G. D., "*Miscible Displacement of Heavy West Sak Crude by Solvents in Slim Tube,*" SPE 18761, Proceedings of 1989 SPE California Regional Meeting, Bakersfield, CA, April 5-7, 1989.

22. Hirshberg, A., deJong, L.N.J., Schipper, B.A. and Meijer, J.G., "Influence of Temperature and Pressure on Asphaltene Flocculation", SPEJ, No. 6, pp 283-293, 1984.
23. Burke, N.E., Hobbs, R.D. and Kashou, S.F., "Measurement and Modeling of Asphaltene Precipitation from LIve Reservoir Systems, "SPE 18273, Proceedings of 63rd Annual Technical Conference and Exhibition of SPE, Reservoir Engineering Volume, Houston, TX, pp 113-126, Oct 2-5, 1988.
24. Coats, K.H., "Equation of STate PVT Program: Users Manual", Scientific Software Intercomp, Denver, Colorado, Nov 1984.

MORPHOLOGICAL SIZE OF ASPHALTENE MICELLES

IN ASPHALT AND HEAVY RESIDUE

Jiunn-Ren Lin, Hsienjen Lian, Jason Chen and Teh Fu Yen

Civil and Environmental Engineering
University of Southern California
Los Angeles, CA 90089-2531

Several techniques have been used to measure the morphological sizes of asphaltene micelles. Among these, the Guinier method for small angle x-ray scattering of both asphalt and heavy oils has been found to be especially useful. From radii of gyration determinations, the morphological diameters of asphaltene micelles are known to be ca. 10 nm. This can be supported by data from dynamic nuclear polarizations as well as from small angle neutron scattering work. Association and clustering of these basic micelles can be chemically or physically initiated by dipole-dipole interactions. The metal shadowing transmittance electron microscopy has been used to demonstrate these phenomena.

INTRODUCTION

Asphalt is defined by ASTM[1] as a dark to black cementitious material in which the predominating constituents are bitumens that occur in nature or are obtained in petroleum processing. Asphaltene occurs in native petroleum and is enriched in asphalt when the light fractions of petroleum are removed during the refining process.

Generally, asphalt can be fractionated into four important fractions: saturates, aromatics, resins, and asphaltenes[2,3]. The classic definition of fractions of asphalts is based on the solution properties of petroleum residuum in various solvents. A complete fractionation scheme is given in Figure 1: (1) the oil constituents are propane soluble; (2) resins are n-pentane soluble but propane insoluble; (3) asphaltenes are toluene soluble but n-pentane insoluble; and (4) preasphaltenes are insoluble both in n-pentane and toluene. The fractionated part of oil is generally considered to be a combination of saturates and aromatics. The polarity of these four fractions increases from saturates, to aromatics, to resins, to asphaltenes.

Asphalt is thought of as a colloidal system similar to petroleum. In the asphalt, it comprises (a) asphaltene, the dispersed phase or micellar phase, (b) resin, the peptizing agent, and (c) gas-oil (saturates and aromatics), the dispersion medium or the intermicellar medium. The minor constituents are organometallics, metals and polar molecules (such as surfactants). Chemically, asphaltene is a very complex macromolecule belonging to the multipolymers, and only a statistically averaged chemical structure is known. Although the asphaltene contents vary in each asphalt, there is no asphalt which does not contain asphaltene. Also, the existence of asphaltene controls the physical properties of asphalt system.

Particle Technology and Surface Phenomena in Minerals and Petroleum
Edited by M.K. Sharma and G.D. Sharma, Plenum Press, New York, 1991

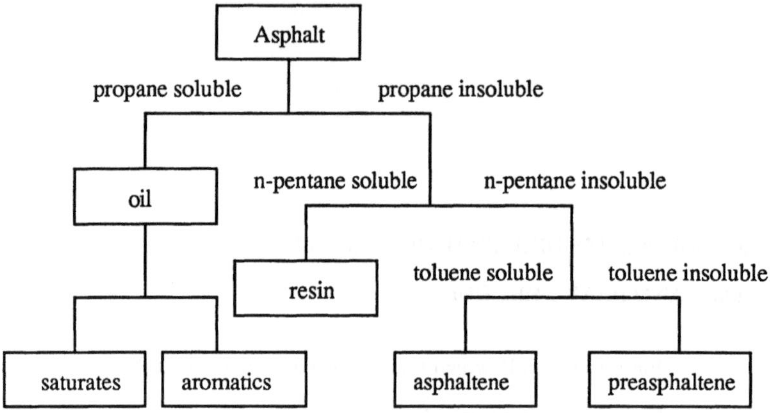

Figure 1 Fractionation scheme of asphalt

Asphalt is a lyophilic colloid in which asphaltene interacts with the dispersion medium through the peptizing agent. Because of the hydrogen bonding and dipole-dipole interaction, the asphaltene particles form micelles and even associations/clusters in the asphalt system.

Structurally, asphaltene contains flat sheets of condensed aromatic systems that may be interconnected by sulfide, ether, aliphatic chains or naphthenic ring linkages. Gaps and holes appear as defect centers in the aromatic systems with heterocyclic atoms coordinated to transition metals such as vanadium and nickel, most likely caused by free radicals. Due to the complexity and the large size of asphaltene molecules, asphaltene particles conveniently fall within the colloidal range. The structure of asphaltene has been determined previously by the x-ray diffraction method[4] and is shown as Figure 2.

Asphaltene dispersed in the asphalt can form micelles, supermicelles, and even giant supermicelles or liquid crystals, depending on the content of asphaltene and resins. All units exist in the asphalt systems but their distributions are different. The micelles and supermicelles are predominantly sol type asphalts; liquid crystal are predominantly gel type asphalts; and giant supermicelles are predominantly sol-gel types. For different types of asphalt, the physical and chemical properties are different; and, therefore, their uses and applications will differ.

EXPERIMENTAL

Eight different asphalt samples belonging to three different colloidal types were used in this study. Asphaltene sample that were used in the TEM metal shadowing were isolated by

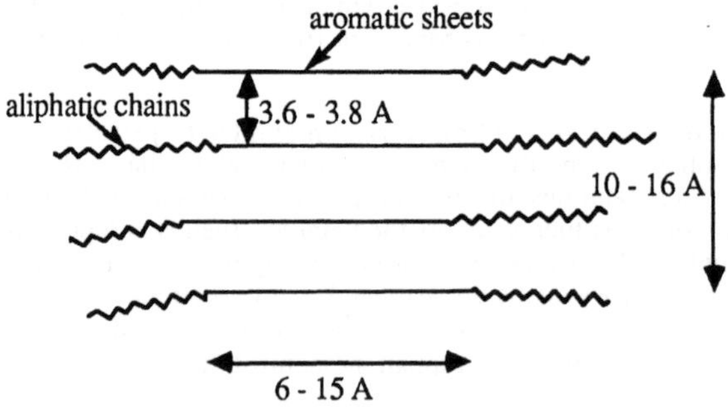

Figure 2 The structure of asphaltene cell

Table I The sources, refinery processes and the type of asphalt samples

Sample	Source	Refinery processes	Types
AAM-1	West Texas	solvent	sol
AAD-1	California Coast	distillation	sol-gel
ABA-1	West Texas Intermediate / West Texas Sour	air-blown	gel
AAA-1	Lloydminister	distillation	sol-gel
AAE-1	Lloydminister	air-blown	gel
AAG-1	California Valley	distillation	sol
AAK-1	Boscan	distillation	gel

Soxhlet process using n-pentane as an eluent for more than 48 hours. The sources and refinery processes of these asphalt samples are listed in Table I. All of the asphalt samples were obtained through the material bank of Strategic Highway Research Program (SHRP) which was operated by University of Texas, Austin.

Small Angle Scattering

Asphalt samples were heated and cooled down before the experiments began. Samples were used immediately after they cooled down and the reminder was discarded after each experiment was finished. Thus, the asphalt samples were free of physical and chemical age hardening.

Experiments were conducted on a Seemens K710H X-ray Generator using 50KV, 20 Ma and monochromatic CuK radiation with an 0.05° slit. Scattering curves of a number of asphalt samples were recorded with a step fraction of 0.02° and ending at 1.00°. The incident beam was a Ge monodiramator with 30 seconds counting time.

Toluene, carbon tetrachloride, and THF were used to disperse the asphalt and 2% by weight samples were prepared. All the solvents used were reagent grade and required no further treatment before their use. Solvent blank tests were done before the samples were tested. All data used to calculate the morphological diameter have been deduced from the background (scattering caused by solvents).

Metal Shadowing by TEM

Four asphaltenes were isolated by Soxhlet process for at least 36 hours using n-pentane solvent as eluent. One AAK-1 oxidized 120 hours asphaltene and one acid fraction of AAK-1 were received from Western Research Institute.

For each run, the asphaltene sample was prepared for a 0.1% (wt/vol) benzene dispersion. A low-velocity nebulizer and high-purity Freon (as the gas phase) were used in the spraying process. Approximately 1 ml of dispersion was sprayed each time through a 90^0 glass tube with 1.0 cm diameter, the exit orifice of which was located 1 foot in front of the SiO surface. For the drying of the samples, the sprayed SiO films were put into an Edwards vacuum evaporator immediately after spraying until vacuum pressure reach 10^{-2} torr. After drying, a 10,000 voltage current was used to shoot Palladium homogeneously on SiO film at an 11^0 angle. The Philips EM 420 microscope was used to perform the research at different magnifications (62,500x, 105,000x, and 230,000x).

RESULTS

Small Angle Scattering

The Guinier treatment has been used widely to measure the dimension of suspended particles[5,6]. Assuming a Gaussian distribution,

$$I = I_0 \exp(-kh^2 R^2) \qquad (1)$$

where I is the scattered intensity, I_0 is scattered intensity for $h = 0$ or $\theta = 0$, R is the scattering mass radius resembling the radius of gyration, and k is a shape constant.

Since the asphaltene micelle is close to a spherical shape

$$k = 1/3$$

Also, in general,

$$h = 4\pi \sin\theta / \lambda = (2\pi \tan 2\theta)/\lambda$$

Thus

$$\log I = \log I_0 - R^2h^2/(3x\ 2.303) \qquad (2)$$

The slope becomes

$$m = -R^2 / 6.909$$

or the radius of gyration

$$R = (-6.909\ m)^{1/2}$$

Therefore, the morphological diameter[7] was calculated by

$$D = 2\ (5/3)^{1/2} R \qquad (3)$$

The values of these calculated radii are summarized in Table II.

In Table II, a number of asphalt dispersions in a variety of solvents are listed. In general, our results indicate that the radii of gyration as evaluated from the Guinier plot is around 4 nm and the morphological diameter is ca. 10 nm. This value does not vary with polarity of the solvent or concentration.

Metal Shadowing by TEM

Table III list the results obtained for different asphaltene samples at different parameters (concentration, magnification, shadow angle). Table IV shows the planar size and stack height of asphaltene samples at different concentrations.

DISCUSSION

Small Angle Scattering

Though the solvent polarity will affect the dispersion and further agglomerate of asphalene micelles, the basic size of the asphaltene micelle will still exist in the entire system. From Table II, it has been seen that regardless of which asphalt, the limiting particle size (micelle size) is similar. This fact confirms the morphological diameters for the asphaltene micelles, which have been studied for over 25 years (Table III). In Table V, all techniques for obtaining these morphological diameters are through the use of the radius of gyration,

Table II The morphological diameter of asphaltenes in different solvents

Sample	Solvent	Radius of gyration (nm)	Morphological diameter (nm)
AAA-1	THF	4.1	10.6
AAK-1	THF	3.9	10.1
AAM-1	toluene	4.0	10.3
	chloroform	4.0	10.3
	THF	4.0	10.3
AAD-1	toluene	4.0	10.3
	chloroform	4.0	10.3
	THF	4.0	10.3
ABA-1	toluene	4.0	10.3
	chloroform	4.0	10.3
	THF	4.0	10.3

either from small angle x-ray scattering (SAXS) or from small angle neutron scattering (SANA). The importance of this is that the diameter of this spherical-shape (or ellipsoidal) micelle is ca. 10 nm. In the studies, asphaltenes dispersed in common organic solvents, as well as crude petroleum or residual oil; all have this basic unit value.

Metal Shadowing by TEM

From the asphaltene structure, the size of the asphaltene cell can be noted as about 2 nm.

Table III The cluster dimension of some asphaltene samples

Sample	Concentration (ppm)	Magnification	Shadow angle(^{0})	Planar size (nm)	Height (nm)
AAA-1	1000	560,000	11	20	2
AAE-1	100	62,500	5.4	36 (W) 241 (L)	4 - 8
AAG-1	1000	105,000	11	107	35
AAK-1	1000	105,000	11	215	9
AAK-O	1000	105,000	11	93	42
AAK-O	1000	230,000	11	80	42
AAK-A	1000	105,000	11	79	20

Table IV The planar size and height of asphaltene samples at different concentration

Sample	Concentration (ppm)	Magnification	Shadow angle(^{0})	Planar size (nm)	Height (nm)
AAK-1	100	105,000	11	62	4
	500	105,000	11	62	7
	1000	105,000	11	215	9
AAK-O	100	105,000	11	72	7
	500	105,000	11	48	11
	1000	105,000	11	93	42
AAK-O	100	230,000	11	90	9
	500	230,000	11	73	11
	1000	230,000	11	80	42
AAK-A	100	105,000	11	64	8
	500	105,000	11	41	11
	1000	105,000	11	79	20

* AAK-O: AAK-1 sample after oxidation; AAK-A: acid fraction of AAK-1 sample

Metal shadowing shows that the height is from 2-42 nm, implying that the phenomena of association did happen in the asphaltene particles.

In Table IV, it is obvious that the particle stack height increases as the asphaltene concentration increases. The polarity increases when the asphalt sample has been oxidized. Due to the polarity increase, attraction between the asphaltene cell will increase,thus increasing the associates. Table III shows that the stack height increases from 9 nm to 42 nm after oxidized at 1000 ppm concentration.

CONCLUSION

From the Table II, it can be found that no matter what type of asphalt, all basic units of asphaltene cells are the same size. No matter what solvent is used, the asphaltene micelle is ca. 10 nm, and is consistent with the results of other methods.

Comparing the asphaltene structure and the data from Table III, the phenomena of association can be seen. The stack height of the cluster will increase according to concentration increase, and the stack height of oxidized asphalt will be larger than that of the original asphalt.

ACKNOWLEDGEMENT

Initial phase of this work was supported by SHRP through NRC. Remaining portion of the work was supported by USC gift account 2086 through Thygard Oil Co. and EER, Inc.

Table V Morphological Diameter of Asphaltene in Petroleum or in Asphalt

Source of crude	Radii of gyration (nm)	Morphological diameter (nm)	Researcher
Lagunillas	3.4	8.8	Dwiggins, 1965
Santa Maria	3.2	8.3	ibid
Tia Juana	4.0	10.3	ibid
Baxterville	4.0	10.3	ibid
Bachaquero	3.1	8.0	ibid
Rhodes	6.9	14.0	ibid
Baxterville	3.4	8.8	Pollack and Yen, 1970
Baxterville	4.0	10.3	ibid
Jobo	2.9	7.5	Dwiggens, 1978
Santa Maria	3.1	8.0	ibid
Safanya	---	13.0	Ravey et al., 1988
Arabian Light	---	12.0	Ravey et al.
Boscan	---	14.0	Ravey et al.
Batiraman	---	14.0	ibid
Cerro-Negro	---	14.0	ibid
Vacuum Resin	5.3	13.7	Overfield et al., 1989
Lyoydminister	4.1	10.6	our work
Boscan	3.9	10.1	our work
West Texas	4.0	10.3	our work
California Coastal	4.0	10.3	our work
West Texas Intermediate / Sour	4.0	10.3	our work

REFERENCES

1. Standards on Petroleum Products and Lubricants (ASTM D8), in: "American Society for Testing and Materials," Philadephia, Pa. (1989).
2. K. H. Altgelt, D. M. Jewell, D. R. Latham, and M. L. Selucky, "Chromatography and Petroleum Analysis," Marcel Dekker Inc, New York and Basel (1976).
3. Standard Test Method for Separation of Asphalt into Four Fractions (ASTM D4124), in: "American Society for Testing and Materials," Philadephia, Pa. (1989).
4. J. T. Kwan and T. F. Yen, Investigation of Structure of Petroleum Asphaltene by X-ray Diffraction, Am. Chem. Soc., Div. Fuel Chem., Preprint 21:67 (1976).
5. C. W. Dwiggins, Jr., A Small Angle X-Ray Scattering Study of the Colloidal Nature of Petroleum, Journal of Physical Chemistry 60(10):3500 (1965).
6. S. S. Pollock and T. F. Yen, Structural Studies of Asphalts by X-ray Small Angle Scattering, Analytical Chemistry 42:623 (1970).

7. O. Kratky, X-ray Small Angle Scattering with Substances of Biological Interest in Diluted Solutions," Prog. Biophy 13:105 (1963).
8. D. H. Kay, "Techniques for Electron Microscopy," F. A. Davis Co., Philadelphia, (1965).
9. J. P. Dickie, M. N. Haller, and T. F. Yen, Electron Microsopic Investigations on the Nature of Petroleum Asphaltics, J. Colloid and Interface Science 29 (1969).

ASPHALTENE PARTICLE SIZE DISTRIBUTION STUDIES BY FRACTALS

Jiunn-Ren Lin, Hsienjen Lian, Kazem M. Sadeghi and Teh Fu Yen

Civil and Environmental Engineering
University of Southern California
Los Angeles, CA 90089-2531

The solubilities of three different asphalt colloidal types (sol, sol-gel, and gel) were determined in various solvents. Asphalt cannot completely dissolve in the solvent, and therefore forms a colloidal system. An unknown number of hydrocarbon particles of different particle size are dispersed in this solution. Korčák distribution has been proposed previously to estimate the size distribution of islands in ocean archipelagos using the concept of fractals; it was employed in this work as an attempt to estimate the size distribution of asphalts. From Korčák distribution, we derived an additional equation to determine the size distribution and particle irregularity in the asphalt colloidal solution. From the relationship between the solvent molecule size and the mass of asphalt precipitated per molar solvent in various solvents, size distribution and irregularity could be obtained by the derived Korčák distribution. It was found that the value of size distribution for the three different colloidal asphalts is different, and it can be used to indicate sol, sol-gel, and gel types of asphalt. In this particular study, however, the values of irregularity were random among the samples, and therefore failed to be a tool to differentiate the three colloidal types.

INTRODUCTION

Asphalt is thought of as a colloidal system similar to petroleum, the difference being that the lighter molecules have been removed from asphalt during the refining process. Asphalt can be fractionated into four important fractions: saturates, aromatics, resins, and asphaltenes; by either the SARA method[1] or the ASTM D4124 process (standard test method for separation of asphalt into four fractions)[2]. The fractionated part of saturates and aromatics is generally considered to be gas-oil. The polarity of these four fractions increases from saturates ---> aromatics ---> resins ---> asphaltenes.

In a colloidal chemistry context, three types of asphalts are recognized: (1) the Newtonian or sol type; (2) the viscoelastic or sol-gel type; and (3) the elastic, non-Newtonian, or gel type[3].

In crude oil, asphaltenes are present as discrete or colloidally dispersed particles in the oily phase. As the various low boiling and intermediate petroleum oils are removed during the distillation process, the particles of asphaltene micelles mass together to form the larger particles. Therefore, the colloidal dispersions (asphalts) are of high molecular weight hydrocarbons (asphaltenes) in a dispersion medium of gas-oil and resins. Though the asphaltenes themselves are insoluble in oil, they can exist as fine or coarse dispersions, depending on the resin content. The resins are part of the oily medium, but have a polarity

Particle Technology and Surface Phenomena in Minerals and Petroleum
Edited by M.K. Sharma and G.D. Sharma, Plenum Press, New York, 1991

higher than gas-oil. This property enables the molecules to be easily adsorbed onto the asphaltene micelles[4].

Air-blowing or air conversion (i.e., oxidation of the asphalt or formation of the gel type of asphalt) involves chemical changes in the asphalt structure and the proportion of the four fractions as the result of this oxidation[5]. No significant change in the percentage of saturates is found, but there is a decrease in both aromatics and resins. Air blowing always results in a significant increase in asphaltenes. Therefore, it may be that the aromatics convert to resins, and resins convert to asphaltenes. As expected, when the amount of total liquid fractions is decreased and that of the solid fractions increased, the softening point increases and the penetration decreases in the resulting composition. Because the physical properties of the three types of asphalt differ significantly from each other, the uses and services of them are also quite different.

Traditionally, there have been many methods proposed to differentiate different types of asphalt, e.g. refinery processes, temperature, susceptibility index, plastic flow index, complex flow index, asphalt aging index, stiffness modulus, and so on. The information that these values can give us is only very rough, and the results for differentiating the types of asphalt is sometimes contradictory and ambiguous. Thus, determination of the type of asphalt generally requires confirmation from several different indexes.

Korčák distribution[6] was proposed to estimate the island size distribution of the archipelago by the concept of fractals. It has demonstrated that the islands distribution satisfy the Korčák distribution:

$$N(A > a) \propto a^{-B}$$

where N(A > a) is the number of islands having an area A greater than a prescribed value a. The value of B is different according to geographic locations[7].

Some researchers have found that the degree of the aggregation of colloidal particles is related to the concentration of asphaltenes and resins in the solution[8-10]. From air-blown and air conversion processes, oxidized asphalt or gel type asphalt has a larger amount of asphaltene than the other two types of asphalt. This implied that for different rheological types of asphalt, the contents of asphaltenes and resins are different, and different degrees of aggregation will occur, thus the size distributions of the three types of asphalt are different from each other. When dissolving the asphalt sample in the alkane-type solvent, a colloidal system forms. In these colloidal systems, a great number of heavy hydrocarbon (asphaltene) particles with different particle sizes will disperse in the medium. We supposed that the distribution of asphaltene particles in the medium was similar to the island distribution of the archipelago and could be described by the derived Korcak law. From the derived Korčák distribution, size distribution value and irregularity value of the asphalt samples would be determined for eight samples (including three different colloidal types of asphalt). The differences for the size distribution values for the three different colloidal types of asphalt (sol, sol-gel, and gel types) can be used as an indicator to differentiate these three colloidal types of asphalt.

EXPERIMENTAL

Eight different asphalt samples from different sites and refining processes were used in this research. All of the asphalt samples were obtained through the material bank of Strategic Highway Research Program (SHRP) which was operated by University of Texas, Austin.

The sources and refinery processes of these asphalt samples are listed in Table I, including the three types of asphalt.

Asphalt samples were heated before the experiments began. The samples were used after cooling to room temperature and the reminder was discarded after each experiment. Thus, the asphalt samples were free of physical and chemical age hardening.

Table I The source, refinery process, and type of asphalt samples

Sample	Source	Refinery processes	Types
AAM-1	West Texas	solvent	sol
AAG-1	California Valley	distillation	sol
AAD-1	California Coast	distillation	sol-gel
AAA-1	Lloydminister	distillation	sol-gel
AAQ-1	Wyoming Canadian Mix	distillation	sol-gel
ABA-1	West Texas Intermediate / West Texas Sour	air-blown	gel
AAK-1	Boscan	distillation	gel
AAE-1	Lloydminister	air-blown	gel

For each run, asphalt sample and solvent were agitated at room temperature for over 48 hours in a flask. The filtration process was done after the agitation. Four solvents, n-pentane, n-hexane, n-heptane, and n-octane, were used in this research. All the solvents used were reagent grade and no further treatment before usage. Whatman No. 1 filter paper were used in the filtration process.

RESULTS

In order to better understand the process for the fractal approach, let us suppose using a divider with an opening r to measure a straight line with length L, the number of steps, N(r), needed will be inversely proportional to r. The length of this line L can be expressed as L= N(r)* r. When the opening r is small enough, the length L can be expressed the same as true length. In the same way, suppose that we try to cover a surface with N balls of radius r. For a smooth surface, $N \propto r^{-2}$. However, for surfaces which remain irregular upon successive magnifications, N grows. As long as high-resolution details are similar to low-resolution features, this growth gives rise to a power law, $N \propto r^{-D}$. The exponent D is a fractal dimension of surface.

If asphalts are dissolved in a quantity of solvent, as mention before, a colloidal system is formed. In the solution, there are a lot of heavy hydrocarbon (asphaltenes) particles with various particle sizes dispersed in the solution. Following the above concept, the distribution of the asphalt solution can be considered according to the Korčák Law:

$$N(V > v) = F\, v^{-D/3} \qquad (1)$$

where N (V > v) = the number of heavy hydrocarbon (asphaltenes) particles whose volume (V) exceed some specific volume (v).
F = a normalizing factor, and
D = a fractal dimension.

The number of particles, dN, with volume between specific v and v+Δv (Δv approachs zero) can be expressed as the following equation:

$$dN = \left(\frac{-D}{3}\right) F\, v^{-\frac{D}{3}-1}\, dV \qquad (2)$$

The total volume of particles in the solution can be expressed as V:

$$V = \int_0^{Vm} \left(\frac{D}{3}\right) F\, v^{-\frac{D}{3}-1}\, dV$$

$$= \frac{DF}{3-D} V_m^{\frac{3-D}{3}} \qquad (3)$$

where Vm = the volume of the largest heavy hydrocarbon particle in solution.

Because the mass of particles (M_T) is equal to the density times the volume of particles, Equation (3) can generate an expression for the total amount of heavy hydrocarbons (asphaltenes) in solution in the system of interest :

$$M_T = \frac{F D \rho}{3 - D} V_m^{(3-D)/3} \qquad (4)$$

where ρ = average density of heavy hydrocarbons (asphaltenes) in the particle.

In order to determine the volume of the largest heavy hydrocarbon particle in solution, we can create another fractal power law to approach[11]:

$$V_m = K' L^{D^*} \qquad (5)$$

where D^* = another dimension, and
L = the characteristic length of the solvent molecule.

Combining Equation (4) & (5), we can get the result as below:

$$M_T = \frac{K F D \rho}{3 - D} L^{D^*(3-D)/3} \qquad (6)$$

where $K = (K')^{(3-D)/3}$.

Table I shows the four different solvents, n-pentane, n-hexane, n-heptane, and n-octane, used to determine asphaltene precipitation for eight asphalts samples AAM-1, AAG-1, AAD-1, AAA-1, AAQ-1, AAK-1, ABA-1, and AAE-1. From these data, we can plot log M_T versus log L to find the slopes and intercepts of the derived Korčák law (equation (6)) for each sample. Figure 1 represents the relationship between the length of the solvent molecule and the amount of asphalt in the solvent for these samples. The intercept and slope of the derived Korčák law equation represents the size distribution and irregularity of asphalt. The size distribution and irregularity of these six samples are shown in Table III and Figure 2.

DISCUSSION

The degree of aggregation of asphaltene is controlled by the content of resin and asphaltene in the solution. Resin, like surfactant, serves as a peptizing agent in the asphalt system and contributes to the dispersion of asphaltene in the solution. Depending on the content of asphaltene and resin, the asphaltene aggregations will differ from each other since there are no two asphalts completely the same. Therefore, the degree of asphaltene aggregation in the asphalt system will not be the same.

It is demonstrated that asphaltene controls the physical properties and performance of the asphalt system. The micelles, giant supermicelles, and liquid crystals are shown dominating

Table II. Yield of precipitates using various solvents for eight asphalt samples.

Sample \ Solvent	Precipitate (wt % asphalt)			
	n - pentane	n - hexane	n - heptane	n - octane
AAA-1	28.1	23.5	21.0	19.5
AAD-1	31.2	28.3	27.3	26.9
AAK-1	33.3	28.1	24.4	24.2
AAG-1	15.0	10.7	9.9	6.6
AAM-1	8.4	5.8	5.5	3.1
AAE-1	34.2	31.7	30.1	27.8
AAQ-1	27.4	24.5	21.7	19.8
ABA-1	31.0	26.3	19.7	18.6

the three types of asphalt. Therefore, the size distribution of asphaltene in the asphalt system should be varied for three different asphalt types. Therefore, it should be valid for asphalt differentiation by the asphaltene size distribution.

Table II shows the asphaltene precipitation for eight asphalt samples with four different solvents. From these data, the value of the mass of asphalt dispersed per molar solvent can be determined and plots log M_T versus log L of the derived Korčák law can be created for each sample. From the slope and intercept, size distribution and irregularity can be obtained.

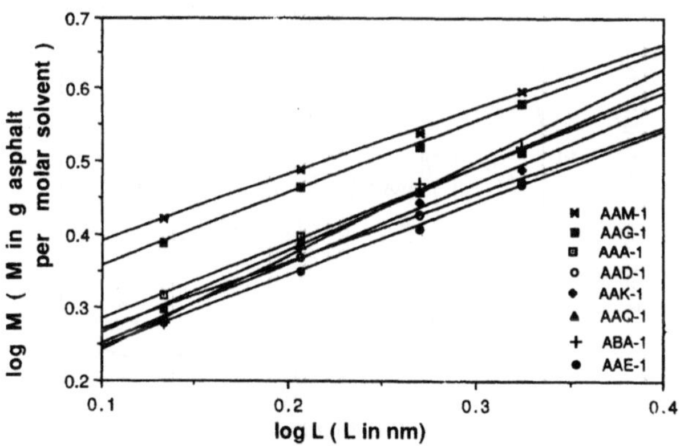

Figure 1 Relationship between the characteristic length of solvent molecule and the amount of asphalt in solvent for gel type asphalt samples

According to Figure 1, the relationship between the length of solvent molecule and the amount of asphalt in the solvent for these samples is obtained. The intercept and slope of the derived Korcak Law equation represents the size distribution and irregularity of asphalt. The results of these eight samples are shown in Table III and Figure 2.

Concentrating in the value of size distribution of Table III and Figure 2, it easy to determine that samples AAM-1 and AAG-1 are in the same group; samples AAA-1, AAQ-1, and AAD-1 are in another group; and Samples AAE-1, ABA-1, and AAK-1 are different from the previous two groups. The values of size distribution for these three groups are around 1.90, 1.51, and 1.39, respectively. These three groups appear to represent three different colloidal types of asphalts, sol, sol-gel, and gel type, respectively.

For the sake of aggregation, the median size of three different colloidal asphalts increased from sol type, to sol-gel type, to gel type. On the same weight basis, the number of particles that can pass through the filter and stay at dispersion will decrease from sol types, to sol-gel types, to gel types of asphalt. Thus, from the derived Korčák Law equation, the size distribution number of the gel type asphalts becomes smaller than the other two types. The results of this research are in agreement with the reduction observed in previous results.

For the reliability consistency test, we are satisfied with consistency, as the standard deviation for the experiment was only 2.43%. The standard deviation of size distribution was only 2.55%. Therefore, it appears that this method is a valid one, although irregularity may not distinguish three different colloidal types of asphalts.

CONCLUSION

The solubility of asphaltene in the solvent with different polarity has been demonstrated by other researchers. According to the polarity of asphaltene and that of various solvents, the degree of asphaltene dispersion is various. The asphaltene size distributions in the different solutions will vary from each other. Using the derived Korčák law, the relationship between the size distribution of asphaltene particles and solvent characteristics are established.

The results presented in Figure 2 and Table II have shown that three catalogies of asphalt can be obtained. These three catagories match exactly with those verified by traditional methods. Since a good reliable consistency could be obtained, the size distribution from Korčák distribution can be applied for the asphalt types differentiation.

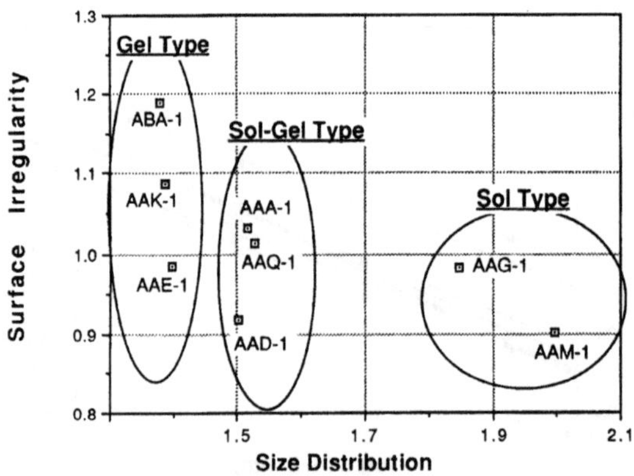

Figure 2 Plot of size distribution vs. irregularity for eight asphalt samples. The three colloidal types of asphalt are indicated in the graph.

Table III. The intercepts (representing size distribution) and slopes (representing surface irregularity) of the derived Korčák law for eight asphalt samples.

Samples \ Distribution	Size distribution $\frac{K F D \rho}{3 - D}$	Surface irregularity $\frac{D^* (3 - D)}{3}$
AAM-1	1.996	0.902
AAG-1	1.848	0.983
AAA-1	1.516	1.031
AAD-1	1.502	0.918
AAQ-1	1.527	1.014
AAK-1	1.387	1.086
AAE-1	1.398	0.986
ABA-1	1.378	1.189

ACKNOWLEDGEMENT

Initial phase of this work was supported by SHRP through NRC. Remaining portion of the work was supported by USC gift account 2086 through Thygard Oil Co. and EER, Inc.

REFERENCE

1. K. H. Altgelt, D. M. Jewell, D. R. Latham, and M. L. Selucky, "Chromatography and Petroleum Analysis," Marcel Dekker Inc, New York and Basel (1976).
2. Standard Test Method for Separation of Asphalt into Four Fractions (ASTM D4124), in: "American Society for Testing and Materials," Philadephia (1989).
3. R. N. J. Saal and J.W. A. Labout, Rheological Properties of Asphaltic Bitumens, Industrial and Engineering Chemistry, 36:823 (1939).
4. T. F. Yen, J. G. Erdman, and S. S. Pollack, Investigation of the Structure of Petroleum Asphaltenes by X-ray Diffraction, Analytical Chemistry, 33:1587, (1961).
5. L. W. Corbett, Asphalt and Bitumen, in: "Ullmann's Encyclopedia of Industrial Chemistry," A3:178 (1985).
6. B. B. Mandelbrot, "The Fractal Geometry of Nature," Freeman, San Francisco (1982).
7. J. Feder, "Fractals," Plenum Press, New York (1988).
8. W. C. Lee, I. Schwager, and T.F. Yen, Determination of the Degree of Association of Coal-Derived Products by Vapor Pressure Osmometry, in: "Chemical and Geochemical Aspects of Fossil Energy Extraction ," Ann Arbor Science, Ann Arbor (1983).
9. C. Aubert and D. S. Cannell, Restructuring of Colloidal Silica Aggregates, The American Physical Society, 56(7):738 (1986).

10. B. M. L. Rao and J. E. Serrano, Viscometric Study of Aggregation Interactions in Heavy Oil, Fuel Science & Technology International, 4(4):483 (1986).
11. D. Avnir and P. Pfeifer, Fractal Dimension in Chemistry : An Intensive Characteristic of Surface Irregularity, Nouveau Journal de Chemie , 7(2):71 (1983).
12. J. R. Lin, H. Lian, K. M. Sadeghi, and T. F. Yen, Asphalt Colloidal Type Differentiated by Korcak Distribution, Fuel, 70, in press (1991).
13. P. J. Crickmore, and C. Hruska, Fractal Geometry, the Korčák Law and Asphaltene Precipitation, Fuel, 68:1488 (1989).

PEPTIZATION STUDIES OF ASPHALTENE IN ASPHALT SYSTEMS AND CORRELATION BY SOLUBILITY PARAMETER SPECTRA

H. J. Lian, J. R. Lin and T. F. Yen

Civil and Environmental Engineering
University of Southern California
Los Angeles, CA 90089-2531

Asphaltene particles are dispersed in saturates and aromatic hydrocarbons (gas-oil) with resins as peptizing agents in asphalt or heavy oil. The interaction between resin and asphaltene micelles is not well understood. In the present study, asphaltene has been dispersed into aromatic hydrocarbons (such as toluene), and the precipitations due to additions of paraffinic hydrocarbons (such as pentane) in the presence of a number of amphiphiles have been studied. These amphiphiles certainly affect the asphaltene precipitation, either by retardation or by enhancement, depending on the structural types and quantities of the amphiphiles. We have found that the nature of resin is that it behaves as an amphiphile, since the polar fractions of resin do contain amphiphiles. The solubility parameter spectra of these asphaltenes are discussed.

INTRODUCTION

Asphalts are classified into natural asphalts and artificial asphalts[1]. Natural asphalts were laid down in geologic strata in pure form (bitumen) or mixed with some mineral material, and artificial asphalts were as a heavy residue from petroleum refineries. Using their different compositions and colloidal characteristics as a guide, asphalts are used in different fields such as paving, roofing, painting and electrical insulation material. In the United States about 70% of all oil asphalts are consumed by the road-paving industry, with some 20% consumed by roofing manufactures.

Because the asphalt system is not a true solution, it can be fractionated into saturates, aromatics, resins, and asphaltenes by the solvent fraction method[2], SARA method[3], or TLC method[4]. The polarity of these four fractions is increased in the order of saturates, aromatics, resins, asphaltenes. In crude oil, asphaltene micelles are present as discrete or dispersed particles in the oily phase. Although the asphaltenes themselves are insoluble in gas-oil (saturates and aromatics), they can exist as fine or coarse dispersions, depending on the resin content. The resins are part of the oily medium but have a polarity higher than gas-oil. This property enables the molecules to be easily adsorbed onto the asphaltene micelles and to act as a peptizing agent of the colloid stabilizer by charge neutralization.

In its natural state, asphaltene exists in an oil-external (Winsor's terminology) or reversed micelle[5] (see Fig. 1). The polar groups are oriented toward the center, which can be water, silica (or clay), or metals (V, Ni, Fe, etc.). The driving force of the polar groups assembled toward the center originates from hydrogen-bonding, charge transfer or even salt formation. This oil-

Particle Technology and Surface Phenomena in Minerals and Petroleum
Edited by M.K. Sharma and G.D. Sharma, Plenum Press, New York, 1991

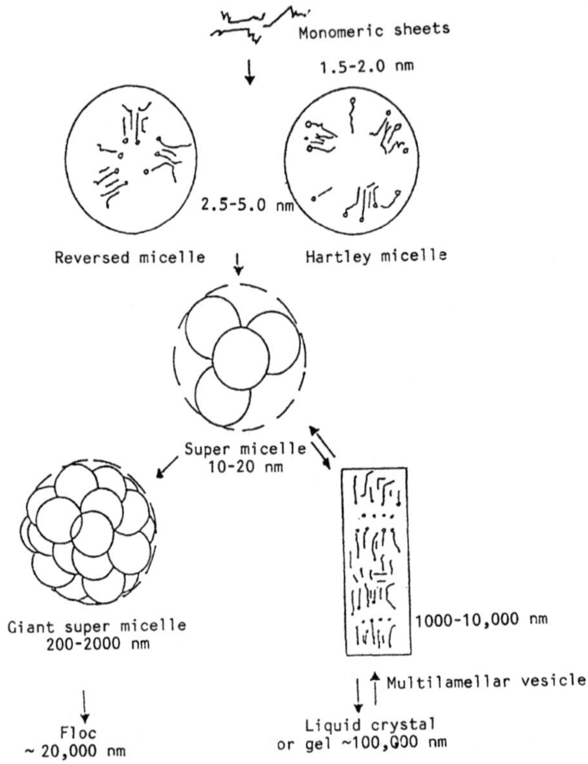

Figure 1. Association, aggregation, and coalescence of micelles to form vesicles and precipitates (flocs). Circle denotes polar functional group, e.g. S, N, and O.

external micelle system can be reversed to oil-internal, water-external micelles (usually called Hartley micelles). An aggregate of asphaltene particles with adsorbed resins can form a supermicelle, and oil may be occluded between supermicelles as an intercellular media. Upon further aggregation the supermicelles can coalesce into giant supermicelles, and can even gradually grow into a liquid crystal.

From the above demonstration, it can be noted that micelle structures are predominant in asphalt with a higher asphaltene content. Three different types of asphalt such as sol (micelle, supermicelle, giant supermicelle), sol-gel (supermicelle, giant supermicelle), gel (liquid crystal) asphalt, can be defined. Most of the paving asphalts belong to the sol-gel type of asphalt, and roofing asphalt belong to the gel (air blown) type of asphalt.

For many years, age hardening (molecular structuring) of paving asphalt has been a major concern in road maintenance. This reversible phenomenon can produce large charges in the flow properties of asphalt without altering the chemical composition of the asphalt molecules. Brown et al.[6,7] studied this reversible molecular structuring (which he called steric hardening) by rheological methods in 1957. Very little work has been conducted in the interim years, however, and no one has approached this topic as a colloidal chemistry problem. Previous studies in Dr. T. F. Yen's group have shown that the structure of the solvents is related to the percentage of liquefaction of a bituminous coal[8] (see Table 1). This article addresses colloidal structure and how it changes with time as controlled by the chemistry of its components, especially the asphaltene to resin ratio. To prove resins are a peptizing agent in the asphalt system, it is important to select a solvent that can easily dissolve asphaltene. For this reason determining the solubility parameter of asphaltenes and resins, as well as a third component (surfactant or emulsifier) that can improve the asphaltene dispersion in the system by the solubility parameter approach, is being attempted. Figure 2 shows the solubility parameters for various crude fractions performed by Yen et al[5].

Table I. Liquefication of a bituminous coal.

Solvent	Molecular Weight	Structural Formula	% Yield
Naphthalene	128	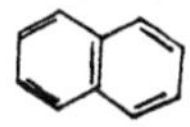	25
Cresol	138	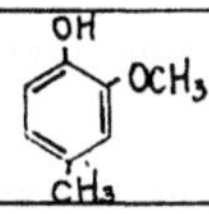	32
Tetralin	132	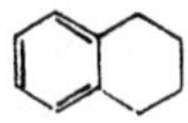	50
O-Cyclohexylphenol	176	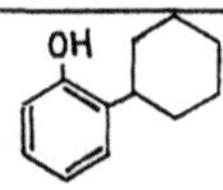	82

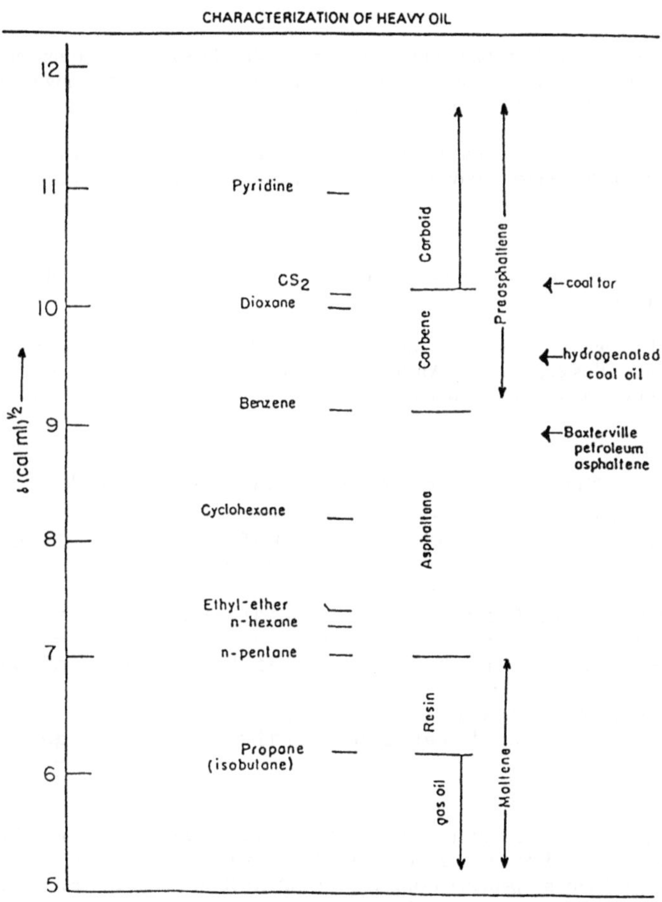

Figure 2. Solubility parameters for various crude fractions.

THEORY AND EXPERIMENTAL

Solubility Parameter

Solubility parameter is defined as the positive square root of cohesive-energy density (potential energy per unit volume):

$$\delta = [\, U^V / V^L]^{1/2}$$

where U^V is the energy change for complete isothermal vaporization of the saturated liquid to the ideal gas state and V^L is the molar volume of the liquid.

For a material to dissolve in a solvent, the free energy change, G_m, of the process must be negative and $G_m = H_m - T\, S_m$. Since the entropy change, S_m, is always positive, the heat of mixing, H_m, determines whether or not dissolution will occur. According to the Hildebrand-Scatchard theory, the heat of mixing given by

$$H_m = V_m\, (\delta_1 - \delta_2)^2 \phi_1 \phi_2$$

where V_m is the total volume of the mixture, ϕ_1, ϕ_2 are the volume fractions, and δ_1, δ_2 are the solubility parameters of the solvent and solute, respectively. Therefore, we can obtain a small or negligible amount of the heat of mixing and make the free energy of change negative only when $\delta_1 = \delta_2$.

There are many methods that can be used to determine the solubility parameter of unknown components. In this study, we estimate the solubility parameter by measuring the solubility in a number of solvents whose δ-values are known[9].

To test solvents with various solubility parameters, the individual solvent as well as the mixed-solvent of the binary mixture are used. The solubility parameter becomes

$$\delta_m = \phi_1 \delta_1 + \phi_2 \delta_2$$

where δ_1, δ_2 are the solubility parameters of the solvents and ϕ_1, ϕ_2 are the volume fractions of the individual solvent.

Table II shows the three different types of asphalt from different sources and refinery processes used in this experiment. All the samples were obtained through the material bank of Strategic Highway Research Program (SHRP), operated by University of Texas, Austin.

Table II. The source and refinery processes of the samples.

Sample	Source	Refinery Processes	Types
AAM-1	West Texas	solvent	sol
AAA-1	Lloydminister	distillation	sol-gel
ABA-1	West Texas Intermediate/West Texas Sour	air blown	gel

All asphalt samples were heated above 120°C. Experiments were performed immediately after the samples cooled down, to avoid physical and chemical age hardening. All asphaltene samples were isolated by pentane using the solvent fraction method[2] (see Fig. 3). All the solvents, including mixtures, used in these experiments were reagent grade, and the solubility parameters of all solvents are listed in Table III.

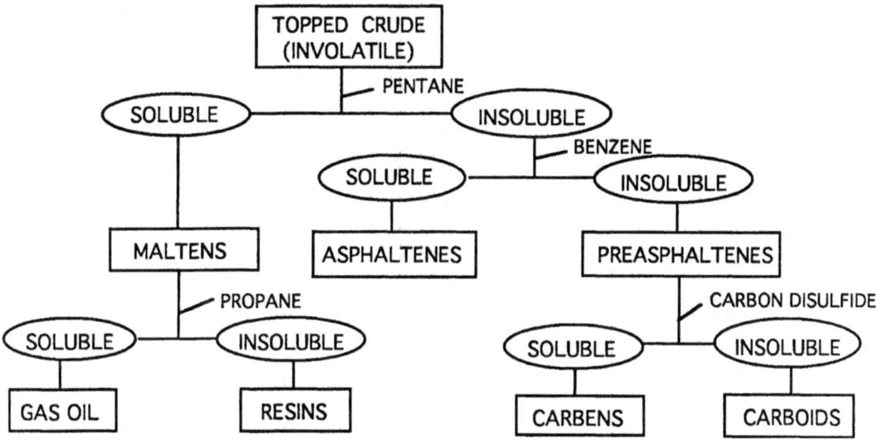

Figure 3. Fractionation and classification scheme for crude oil fractions.

Table III. Solubility parameters for solvent/solvent mixture.

Solvent / Mixture	Solubility Parameter (δ)
n-pentane	7.0
n-hexane	7.4
n-pentane / cyclohexane	7.8
cyclohexane	8.2
carbon tetrachloride	8.6
toluene	8.9
chloroform	9.4
carbon disulfide	10.1
carbon disulfide / pyridine	10.5
pyridine	10.9
carbon disulfide / butanol	11.0
pyridine / butanol	11.2
butanol	11.3

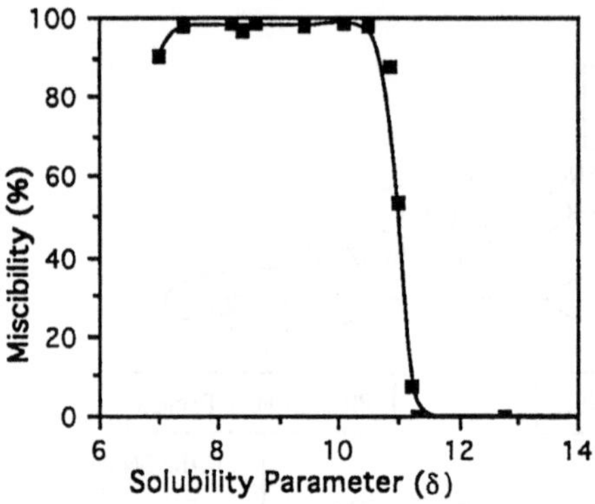

Figure 4. The parameter spectrum for sample AAM-1 asphalt

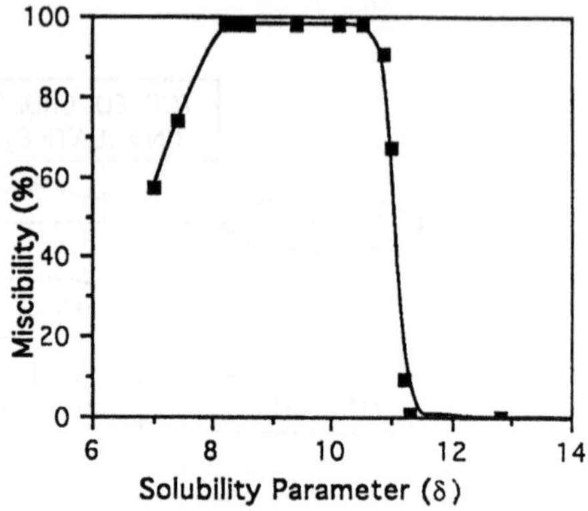

Figure 5. The parameter spectrum for sample AAA-1 asphalt

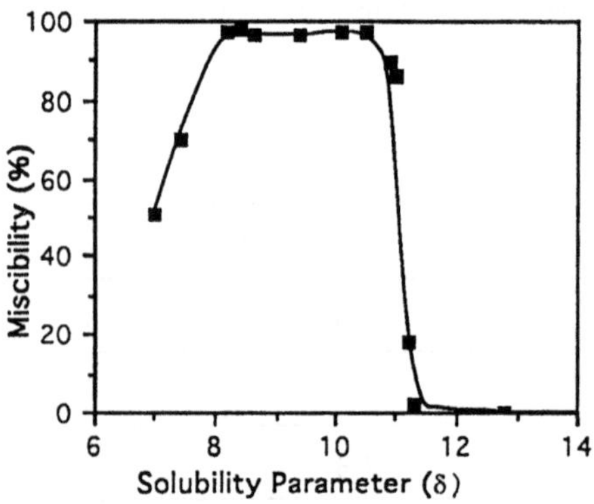

Figure 6. The parameter spectrum for sample ABA-1 asphalt

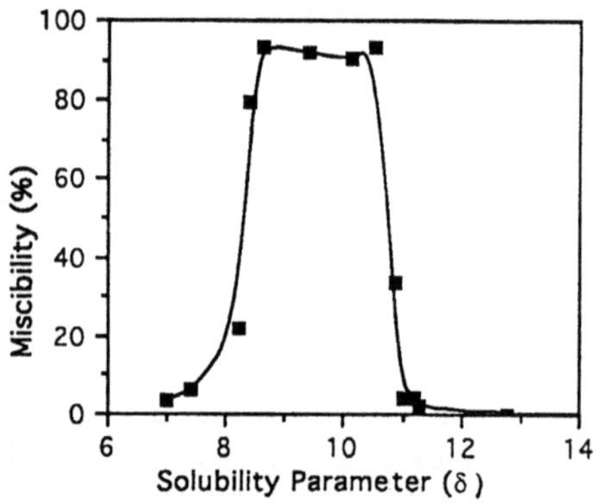

Figure 7. The parameter spectra for sample AAM-1 asphaltene

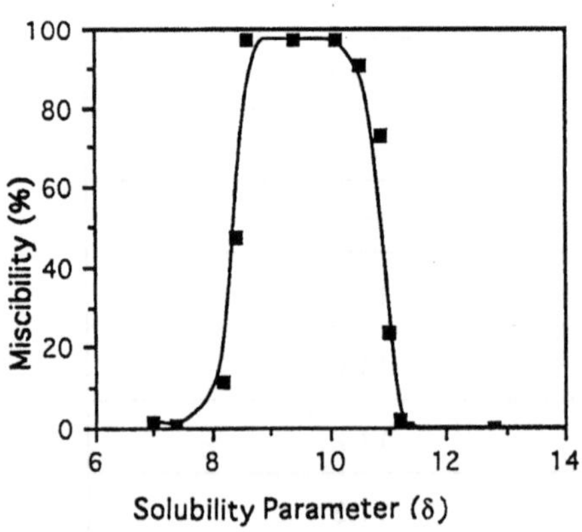

Figure 8. The parameter spectrum for sample AAA-1 asphaltene

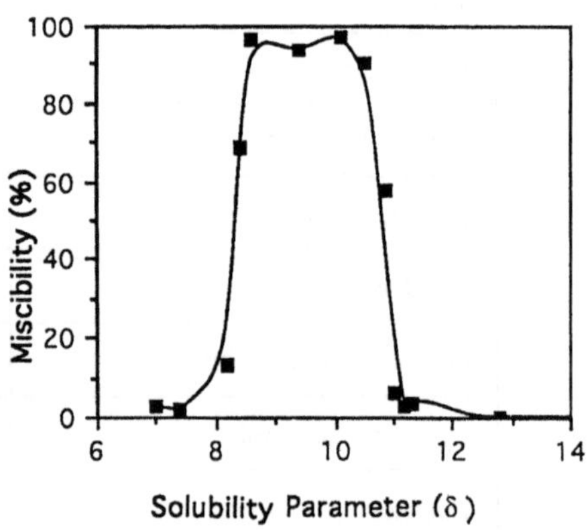

Figure 9. The parameter spectrum for sample ABA-1 asphaltene

For each run, 0.5 gm of asphalt or asphaltene with 10 ml solvent were placed in a flask, then agitated by a magnetic stir bar for 30 minutes at room temperature. Finally, precipitation of solutes was filtered out by Whatman No.1 filter paper. The solubility parameter of asphalt or asphaltene was determined by miscibility (compatibility).

Peptization Test

First, traditional solvent methods were used to remove soluble impurities from precipitates (asphaltenes and preasphaltenes). The precipitates were then dissolved in toluene solution to remove the precipitated preasphaltenes. Finally, relatively pure asphaltenes were re-dissolved in toluene to obtain a 100 ppm concentration of asphaltene solution[10].

The formula and properties of the various amphiphiles assayed as peptizing agents are presented in Table IV in which resins were isolated by the TLC method. Table V presents the property and formula of toluene which was used as a solvent in this experiment. Asphaltene precipitation from toluene solutions was tested by adding pentane, and was carried out using 100 ppm asphaltene solution containing 0.5% (by weight) nonyl phenol, stearic acid, or Hexadecylamine. After 20 minutes of agitation the solutions were left at room temperature for 3 hours and, afterwards, centrifuged at 3000 rpm for 30 minutes. The absorbance of supernatant was determined at 400 nm by a double beam Varian UV-visible spectrophotometer. Different concentrations (0.5% and 1%) of nonyl phenol were then used to repeat the peptization test following the same procedure. Finally, two different resins (AAA-1, and AAX-1) were compared with nonyl phenol at 50 ppm concentration of amphiphiles, to perform the peptization test. All the above experiments were performed with a 100 ppm asphaltene solution (in toluene) with different kinds of amphiphiles.

Table IV. Amphiphiles assayed for the peptization of asphaltenes in pentane.

Compound	Molecular Weight	Structural Formula
Nonyl Phenol	220	$CH_3(CH_2)_8$-C$_6$H$_4$-OH
Stearic Acid	284	$CH_3(CH_2)_{16}COOH$
Hexadecylamine	241	$CH_3(CH_2)_{15}NH_2$
Resins	800 - 1300	? may include aromatic group & hydroxyl group

Table V. Solvent used in peptization test

Compound	Molecular Weight	Structural Formula
Toluene	92	C$_6$H$_5$-CH_3

RESULTS

We obtained solubility parameter spectra of asphalt and asphaltenes for three different colloidal types of asphalt (AAM-1, AAA-1, and ABA-1). Figures 4-6 represent the solubility parameter spectra for asphalt, and Figures 7-9 represent that for asphaltenes. Comparing the solubility parameter of asphalt and asphaltenes, it can be found that the solubility parameters of asphalt and asphaltenes are around 7-10.5 and 8.5-10, respectively. Without a doubt, the composition of asphalt is more complex than that of asphaltenes. Therefore, the solubility parameter of asphalt has a wider range than that of asphaltenes.

From Figures 4-6, we can easily observe significant differences for three different types of asphalts. Comparing Figures 7 to 9, all spectra for the three samples are similar, except for the tail part. This may indicate that the difference in the composition of asphaltenes in the three asphalt types is not significant. It is demonstrated that the compositions for the three different types of asphalt differ significantly due to the existing different compositions of asphaltenes which do not dissolve well in solvents.

Figure 10 shows that adding 0.5% by weight of nonyl phenol, stearic acid, and hexadecylamine exhibits different concentrations of asphaltenes in solution when pentane volumes are above 50%. At a 70% volume of pentane, the concentration of asphaltenes in 0.5% hexadecylamine solution was less than the standard solution (without amphiphiles). The difference was about 6.2 ppm, meaning that hexadecylamine could be a flocculation agent. The efficiency of different amphiphiles for peptization tests decreases in the order of nonyl phenol, stearic acid, standard solution, hexadecylamine. No doubt, nonyl phenol is the best peptizing agent tested in this experiment.

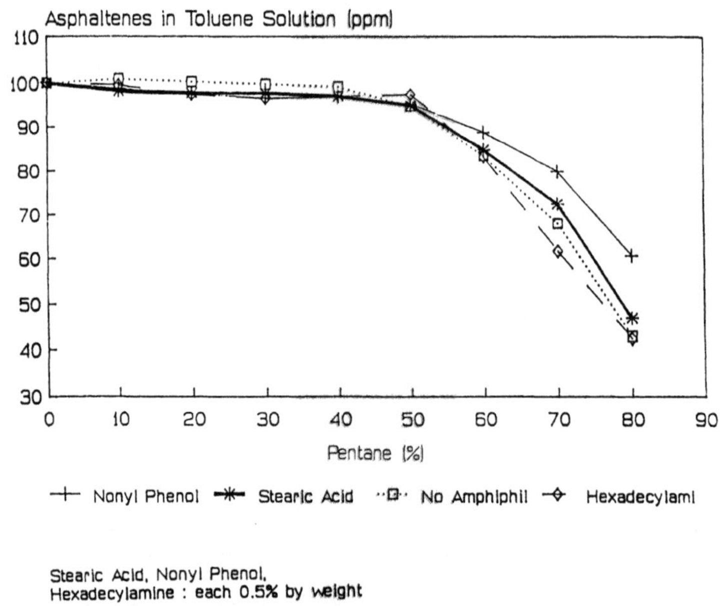

Figure 10. The comparison of different amphiphile on the precipitation of AAA-1 asphaltenes by pentane.

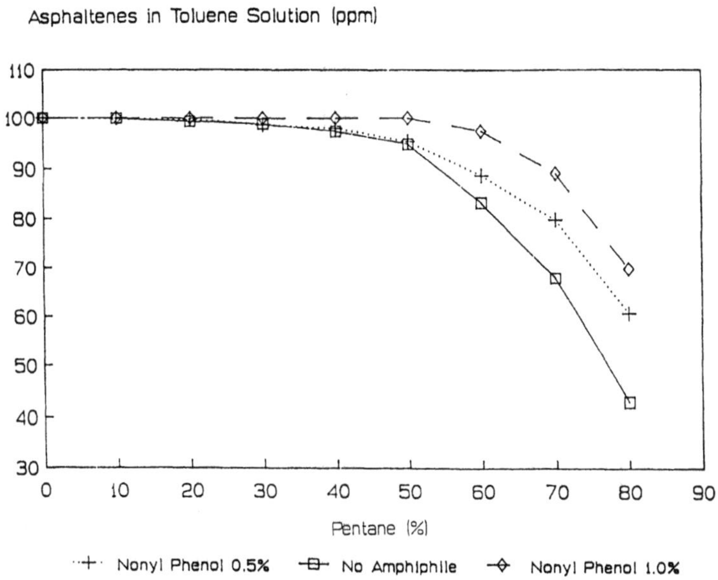

Figure 11. The comparison of different concentration of nonyl phenol on the precipitation of AAA-1 asphaltenes by pentane.

In order to correlate peptizing efficiency with the concentration of amphiphiles, nonyl phenol was tested at different concentrations such as 0.5%, and 1% by weight. The results are shown in Figure 11. The precipitation of asphaltenes at 60% volume of pentane in solution has a 15 ppm difference between 1.0% nonyl phenol solution and standard solution, and the difference increases to 27 ppm when pentane volume reaches 80% in solution. It is clear that the peptizing efficiency increases with the increasing concentration of amphiphiles.

Next, two different resins (AAA-1, AAX-1) were compared with nonyl phenol to determine which amphiphile is the best peptizing agent. From Figure 12 the peptizing efficiency of resins is seen to be better than nonyl phenol. Although AAA-1 resin seems to have a higher peptizing efficiency than AAX-1 resin, those two curves are very close.

CONCLUSION

From solubility parameter experiments, we find that the asphalt system is a colloidal types and has a wider range of solubility parameters. The different compositions of asphaltenes in asphalts have different miscibility in solvents because asphaltenes do not dissolve well in solvents. From peptization experiments, we find that amphiphiles with a molecular weight of at least 220 gm (e.g. nonyl phenol) can be adsorbed into the asphaltene molecule as a peptizing agent. The results indicate that the interactions are not restricted to the polar groups, but the pi electrons of the aromatic and naphthenic portions in the asphaltenes may act as electron donors for hydrogen bonds with hydroxyl groups of the amphiphiles. Resins have proven to be the best peptizing agent in asphalt colloidal systems due to their high molecular weight and high aromatic portion and hydroxyl group. Synthetic asphalts can be made by adding resin fractions which were isolated from original asphalts to solve the age hardening problem. Through this method we can lengthen the paving asphalt life and recover more oil from oil reservoirs.

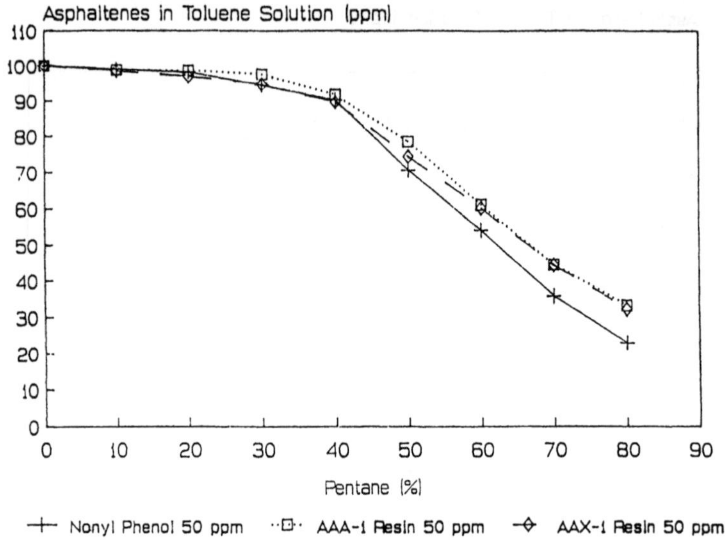

Figure 12. The comparison of adding different amphiphiles on the precipitation of AAA-1 asphaltenes by pentane.

ACKNOWLEDGEMENT

Initial phase of this work was supported by SHRP through NRC. Remaining portion of the work was supported by USC gift account 2086 through Thygard Oil Co. and EER, Inc.

REFERENCE

1. E. J. Barth, "Asphalt Science and Technology", Gordon and Breach Science Publishers, New York (1962).
2. J. Schwager and T. F. Yen,Coal-liquefaction Products from Major Demonstration Processes. 1. Separation and Analysis, Fuel 57:100 (1978).
3. K. H. Altgelt, D. M. Jewell, D.R. Latham, and M. L. Selucky, Chromatography in petroleum analysis, in: "In Chromatographic Science Series," K. H. Algelt and T., ed., Marcel Dekker Inc., New York and Basel (1979).
4. Y. Y. Wang and T. F. Yen, Application of Thin-Layer Chromatographic Identification of heavy Fractions of Fossil Fuels, ACS Pacific Conference on Chemistry and Spectroscopy, Pasadena (1989).
5. T. F. Yen, Asphaltic Materials, in: "Encyclopedia of Polymer Science and Engineering," M. Grayson and J. I. Krochwitz, ed., Wiley, New York (1988).
6. A. B. Brown, J. W. Sparks, and F. M. Smith, Steric Hardening of Asphalts, Proc. Assoc. Asp. Pav. Technol. 36:486-494 (1957).
7. A. B. Brown, J. W. Sparks, and F. M. Smith, Visco-elastic Properties of a High Consistance Asphalt, J. Colloid Sci. 12:283-293 (1957).
8. T. F. Yen, The Nature of Asphaltenes In Heavy Oil, Energy Sources 1:321-326 (1974).
9. V. A. Weinberg, J. I. White, and T. F. Yen, A study of the Solvent Fractionation of Petroleum Pitch for Mesophase Formation, Fuel 62:1903-1509 (1983).
10. E. Wong and T. F. Yen, Formation of the Carbonaceous Mesophase from a Petroleum Derived Paving Asphalt, Energy Sources 10:201-208 (1988).

EMPIRICAL EXPRESSION OF PERMEABILITY IN TERMS OF OTHER PETROPHYSICAL PROPERTIES

George V. Chilingarian

School of Engineering
University of Southern California
Los Angeles, California, USA 90089–1211

ABSTRACT

Using the multi-variable linear regression analysis, the author developed empirical expressions for permeability in terms of porosity, specific surface area, and irreducible fluid saturation for four carbonate reservoir rock areas in the USSR. The coefficient of correlation varied from 0.981 to 0.997.

INTRODUCTION

Permeability, which characterizes the ability of rocks to allow the movement of fluids contained in their pores, is one of the most important parameters describing the porous media. Normally, in order to measure the permeability, the sample must have a simple geometric shape (e.g. cylinder or cube) and certain dimensions. On the other hand, measurements of porosity, pore-size distribution, and specific surface area do not require special geometric dimensions. The correlations among permeability and other "easier-to-measure" parameters, therefore, have been studied theoretically and experimentally[1]. In practice, the most often reported correlation is that between the permeability and porosity[2, 3]. The coefficient of correlation for porosity–permeability relationship varies from sample to sample, with a better correlation if the porosity used in the calculation is measured when a core contains the irreducible fluid. Porosity does not reflect the number and width of fractures, the pore sizes and topological structure, whereas the specific surface area does. Thus, it appears advisable to relate permeability simultaneously to porosity, specific surface area, irreducible water/oil saturation, grain size/pore size/throat size distribution, tortuosity, etc.

In this study, the author examined the interrelationship among permeability k, porosity ϕ, specific surface area s_s, and residual water saturation S_{wr} for carbonate reservoirs of USSR[1]. Parameters influencing the permeability most were included, whereas unnecessary ones were not used.

Particle Technology and Surface Phenomena in Minerals and Petroleum
Edited by M.K. Sharma and G.D. Sharma, Plenum Press, New York, 1991

FUNCTION FORM

The following form of mathematical relation was used by the writer:

$$\log k = a_0 + a_1 S_{wr} + a_2 s_s + a_3 \phi + a_4 S_{wr} \cdot s_s + a_5 S_{wr} \cdot \phi + a_6 s_s \cdot \phi. \quad (1)$$

The above equation contains three linear terms and three cross-product terms. This equation was used by the writer because of its simplicity—it can be linearized as follows

$$Y = \sum_{j=0}^{m} a_j X_j \quad (2)$$

where $Y = \log k$ and x_j's are the corresponding terms following a_j in Eq. 1; m is equal to 6 here.

The constants a_0, a_1, ..., a_m are estimated by a standard linear regression technique[5, 6], i.e., by minimizing:

$$Q(a) = \sum_{i=1}^{N} \left(\log k_i - \sum_{j=0}^{m} a_j X_{ij} \right)^2 \quad (3)$$

where Q is the squared difference between the measured and calculated values.

CASE STUDIES

Data used in this study represent four carbonate reservoir areas in the USSR. The first one examined is Vuktyl'skiy gas-condensate deposit in the USSR (Table I).

All the six terms (see Eq. 1) were used to calculate the regression equation, and the value of 0.9967 for the correlation coefficient R indicates that a strong relation exists between permeability and other petrophysical properties (Table II). The porosity used here is the "intercommunicating" porosity measured when the pores do *not* contain irreducible fluids, which usually gives a poor correlation between permeability and porosity alone[2]. This is because dead–end pore spaces, which hold part of the irreducible fluid, contribute little to the fluid flow. Because irreducible water saturation is included in the equation, however, the dead-end pore spaces have been indirectly considered. Specific surface area, together with the irreducible water saturation, reflects the pore/grain size of the rock sample, which can significantly result in totally different values of permeability for two rock samples which may have the same values of porosity. Specific surface area also takes into account the presence of fractures, which greatly increase the permeability with only a minor effect on porosity (porosity due to the fractures alone $\leq 1\%$).

Next, the contribution of each term was examined and unnecessary ones, having negligible effect on permeability, were removed. This is achieved by using the F test. In Eq. 1, each term has a corresponding F-value. The larger the F-value, the stronger the corresponding term affects the permeability. The terms with F-values

Table I. Petrophysical properties of carbonate rocks of Vuktyl'skiy gas condensate deposit. (After Bagrintseva[4], table 37, p 151.) Lithology varies from dolomites (porous, porous–cavernous, finely-cavernous, crystalline, slightly calcareous, crystalline dense, etc.) to true limestones (dolomitized, with very fine caverns, and micro-grained).

S_{wr},%	s_s, cm^2/cm^3	ϕ^*, %	$S_{wr}s_s$	$S_{wr}\phi$	$s_s\phi$	k (md)
16	2156	7	34496	112	15092	3.6
28	7070	8	197960	224	56560	0.4
19	3878	8	73682	152	31024	1.6
25	2058	10	51450	250	20580	11.5
12	1827	12	21924	144	21924	26
12	1113	12	13356	144	13356	76
9	1421	13	12789	117	18473	58
9	1428	13	12852	117	18564	63
28	2030	13	56840	364	26390	28
5	945	14	4725	70	13230	138
9	3668	14	33012	126	51352	9.5
7	644	14	4508	98	9016	294
3	854	15	2562	45	12810	208
16	2142	15	34272	240	32130	36
8	1001	15	8008	120	15015	167
4	532	18	2128	72	9576	1011
4	441	20	1764	80	8820	1910

* "Effective" porosity as used in USA; "open" porosity as used in USSR.

Table II. Regression coefficient a_j's and corresponding ratios of variances F_j's. Vuktyl'skiy gas condensate deposit, USSR.

j	fitted coefficient a_j	term	F_j
0	8.6323×10^{-1}	constant	4.21
1	-2.3047×10^{-2}	S_{wr}	0.53
2	-5.3394×10^{-4}	s_s	3.60
3	1.3847×10^{-1}	ϕ	33.48
4	1.0987×10^{-5}	$S_{wr}s_s$	5.58
5	-2.6844×10^{-4}	$S_{wr}\phi$	0.01
6	-1.3178×10^{-6}	$s_s\phi$	0.01
		total	195.17

$F_{0.25}(6, 10) = 1.58$, $F_{0.25}(1, 10) = 1.49$, $R = 0.9967$

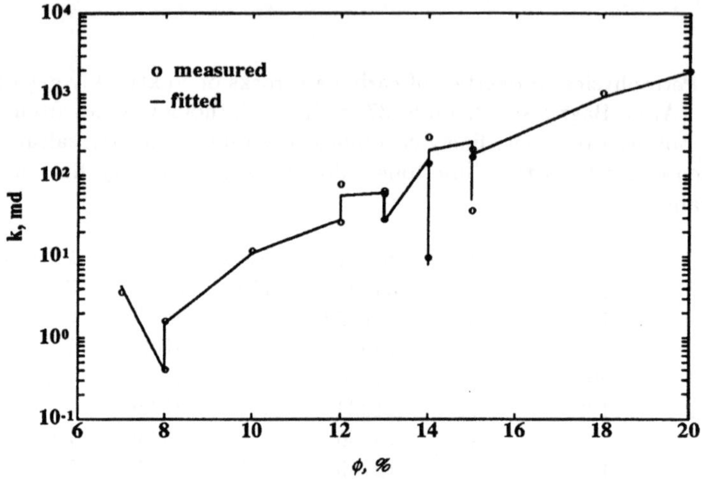

Figure 1. Plots of permeability, k, versus intercommunicating (open) porosity, ϕ_o, where the effects of specific surface area (per unit of pore volume) and irreducible water saturation, S_{wr}, are considered in the data fitting, showing both the measured and fitted data for Vuktyl'skiy gas–condensate deposit, USSR.

smaller than a given critical value are statistically not important in the equation. Only one term with the least F-value, however, should be eliminated each time, after which the constants a_j's must be recalculated. This process is repeated until all the terms have F-values higher than the given critical value. Then, one can state that statistically the remaining terms affect the permeability.

The constant vector (regression coefficient), **a**, is recalculated as shown in Table III. After eliminating two insignificant terms, the matching is as good as the previous one, because the coefficient of correlation remains unchanged. The final

Table III. Regression coefficient a_j's and corresponding ratios of variances F_j's. Vuktyl'skiy gas condensate deposit, USSR.

j	fitted coefficient a_j	term	F_j
0	9.5325×10^{-1}	constant	22.99
1	-2.7880×10^{-2}	S_{wr}	26.46
2	-5.5967×10^{-4}	s_s	139.83
3	1.3309×10^{-1}	ϕ	164.93
4	1.1709×10^{-5}	$S_{wr} s_s$	40.76
		total	331.49

$F_{0.01}(4, 12) = 5.41$, $F_{0.01}(1, 12) = 9.33$, $R = 0.9967$

Table IV. Petrophysical properties of carbonate reservoirs (both limestones and dolomites) of Central Asia, USSR. (After Bagrintseva[4], table 46, p. 177.)

S_{wr},%	s_s, cm^2/cm^3	ϕ, %	$S_{wr}s_s$	$S_{wr}\phi$	$s_s\phi$	k (md)
32	3465	13	110880	416	45045	9.6
16	658	14	10528	224	9212	302
10	231	14	2310	140	3234	2400
25	2324	15	58100	375	34860	29
30	1750	15	52500	450	26250	55
30	3311	16	99330	480	52976	18
25	1169	16	29225	400	18704	151
63	13860	17	873180	1071	235620	1.3
22	1316	17	28952	374	22372	129
58	12880	19	747040	1102	244720	2.1
43	6650	19	285950	817	126350	7.4
37	8358	20	309246	740	167160	5.6
46	6300	22	289800	1012	138600	14
27	3542	22	95634	594	77924	44
23	1848	22	42504	506	40656	150
22	840	23	18480	506	19320	883

correlation equation is as follows:

$$\begin{aligned}\log k &= 0.9532 - 2.7880 \times 10^{-2} S_{wr} - 5.5597 \times 10^{-4} s_s + \\ &\quad +1.3309 \times 10^{-1}\phi + 1.1707 \times 10^{-5} S_{wr} \cdot s_s, \ (R = 0.997) \qquad (4)\end{aligned}$$

The results are plotted in Figure 1. Similarly determined, for carbonate reservoir rocks of Central Asia, USSR (see Table IV), the correlation equation is:

$$\begin{aligned}\log k &= 3.8690 - 1.0536 \times 10^{-1} S_{wr} - 4.1979 \times 10^{-4} s_s + \\ &\quad +6.5363 \times 10^{-6} S_{wr} \cdot s_s + 2.8324 S_{wr} \cdot \phi, \ (R = 0.985). \qquad (5)\end{aligned}$$

For carbonate reservoirs of Kuybyshev Along-Volga Region (see Table V) the correlation equation is:

$$\begin{aligned}\log k &= 2.1085 - 5.0777 \times 10^{-2} S_{wr} - 4.3785 \times 10^{-4} s_s + \\ &\quad +7.9959 \times 10^{-2}\phi + 7.6326 \times 10^{-6} S_{wr} \cdot s_s, \ (R = 0.988). \qquad (6)\end{aligned}$$

The correlation equation for the carbonate reservoir rocks of Orenburg Deposit, USSR (see Table VI) is as follows:

$$\begin{aligned}\log k &= 3.4351 - 2.0443 \times 10^{-1} S_{wr} + 9.5086 \times 10^{-6} S_{wr} \cdot s_s + \\ &\quad +8.0217 \times 10^{-3} S_{wr} \cdot \phi - 2.3892 \times 10^{-5} s_s \cdot \phi, \ (R = 0.981). \qquad (7)\end{aligned}$$

Table V. Regression coefficients a_j's and corresponding ratios of variances F_j's. Carbonate reservoirs of Central Asia, USSR.

j	fitted coefficient a_j	term	F_j
0	3.8690	constant	293.06
1	-1.0536×10^{-1}	S_{wr}	48.07
2	-4.1979×10^{-4}	s_s	37.54
4	6.5363×10^{-6}	$S_{wr} s_s$	45.62
5	2.8324×10^{-3}	$S_{wr} \phi$	25.34
		total	66.31

$F_{0.01}(4, 11) = 5.67,\ F_{0.01}(1, 11) = 9.65,\ R = 0.9853$

Table VI. Petrophysical properties of carbonate reservoirs of the Kuybyshev, Along-Volga Region, USSR. (After Bagrintseva[4], table 32, p. 127.) Lithology is mainly limestone.

S_{wr},%	s_s, cm²/cm³	ϕ, %	$S_{wr} s_s$	$S_{wr} \phi$	$s_s \phi$	k (md)
58	7875	10.1	456750	585.8	79537.5	0.8
7	427	13	2989	91	5551	577
22	2016	14.8	44352	325.6	29836.8	39
21	1890	17.5	39690	367.5	33075	74
16	1330	17.6	21280	281.6	23408	151
41	5285	18.1	216685	742.1	95658.5	10.4
52	10850	19.4	564200	1008.8	210490	3
15	1309	19.4	19635	291	25394.6	209
16	1253	19.7	20048	315.2	24684.1	239
17	700	20.5	11900	348.5	14350	860
42	9520	21	399840	882	199920	5
10	518	21	5180	210	10878	1671
52	7210	23.4	374920	1216.8	168714	11.9
32	5432	23.7	173824	758.4	128738.4	22
18	2310	23.7	41580	426.6	54747	123
14	749	24	10486	336	17976	1207

As one can see, in general, there is a good correlation between permeability and other petrophysical properties. For different lithologies, however, each term weighs differently. The complex pore structures may contribute to such diversity. Care, therefore, should be taken when using such empirical correlations because many factors, e.g, mineralogy and diagenesis can greatly affect the pore configuration.

CONCLUSIONS

The use of multi-variable linear regression analysis enabled the author to demonstrate experimentally that the permeability can accurately be estimated from the other petrophysical properties. It was shown that besides porosity, specific surface area and irreducible water saturation are important factors for such permeability estimations. Such experimental relations should be established for other reservoir rocks.

ACKNOWLEDGEMENTS

The writer is greatly indebted to Jincai Chang whose help was invaluable in carrying this "old dream" to fruition.

REFERENCES

1. Langnes, G. L., Robertson, J. O., Jr. and Chilingar, G. V., 1972. *Secondary Recovery and Carbonate Reservoirs.* Elsevier, New York, 304 pp.

2. Chilingar, G. V., 1964. Relationship between porosity, permeability and grain-size distribution of sands and sandstones. In L. M. J. U. van Straaten (ed.), *Deltaic and Shallow Marine Deposits.* Elsevier, Amsterdam, pp 71–75.

3. Lucia, F. J., 1981. Petrophysical parameters estimated from visual descriptions of carbonate rocks. A field classification of carbonate pore space. SPE of AIME 56th Annual Fall Tech. Conf., San Antonio, Oct. 5–7, SPE 10073: 6pp.

4. Bagrintseva, K. I., 1977. *Carbonate Rocks—Oil and Gas Reservoirs.* Nedra, Moscow, 231 pp.

5. Mao Sisong, Ding Yuan, Zhou Jixiang and Lu Naigang, 1981. *Regression Analysis and Experiment Design.* East China Normal University, 375 pp.

6. Rawlings, J. O., 1988. *Applied Regression Analysis: A Research Tool.* Wadsworth and Brooks/Cole Advanced Books and Software, Pacific Grove, California, 553 pp.

ON RELIABILITY OF DESCRIPTION AND PERFORMANCE DATA ESTIMATES FROM RESERVOIR SIMULATORS

Rufus O. Elemo

Dept. of Engineering Science
Petroleum Engineering Div.
North Dakota State University
Fargo, North Dakota 58105

This paper presents the results of a study to investigate and establish the reliability of both the description and performance data estimates from numerical reservoir simulators. Using optimal control theory, an algorithm was developed to perform automated matching of field observed data and reservoir simulator calculated data, thereby estimating reservoir parameters such as permeability and porosity. Well known statistical and probability methods were then used to establish individual confidence limits as well as joint confidence regions for the parameter estimates and the simulator predicted performance data. The results indicated that some reservoir input data can be reliably estimated from numerical reservoir simulators. Reliability was found to be inversely related to the number of unknown parameters in the model and the level of measurement error in the matched field observed data.

INTRODUCTION

The use of petroleum reservoir simulation as a powerful reservoir engineering tool is well documented in technical literature (3,4,12,14,16). Numerical reservoir simulation is often used for petroleum reservoir characterization and prediction of future reservoir performance under various recovery schemes, well configurations or operating conditions. A major difficulty usually encountered in applying reservoir simulators is lack of adequate and accurate reservoir description data. Mathematical modeling of most petroleum reservoirs' performance requires large number of description data, even when zonation technique is used (3,8). The conventional well test and interpretation methods, based on the exponential integral solution of the reservoir governing equation, can be used to estimate some of the reservoir description data (7). But by applying the technique of history matching, reservoir simulation provides a more accurate means of estimating the initial values of pertinent reservoir description data, particularly when the reservoirs are stress-sensitive (9,11). History matching is a parameter estimation technique that can consider such factors as reservoir hetero-

Particle Technology and Surface Phenomena in Minerals and Petroleum
Edited by M.K. Sharma and G.D. Sharma, Plenum Press, New York, 1991

geneity, variable reservoir thickness, variable fluid properties, gravity forces and other complicating factors in estimating the reservoir description data.

In previous studies, it has been shown that the temporal response of pressure in normally pressured, non-compacting reservoir is sensitive to variations in the formation permeability and porosity, net pay thickness and the specific productivity index while the temporal response of pressure in compacting, geopressured-geothermal reservoir is sensitive to variations in the formation permeability, uniaxial compaction coefficient, net pay thickness and the specific productivity index, but virtually insensitive to variations in the reservoir porosity (5,10). Therefore, using reservoir pressure as the performance data for matching, the technique of history matching has been used to estimate parameters such as permeability and porosity for both compacting and non-compacting reservoirs (2,11,17). History matching technique has been applied to obtain improved reservoir descriptions for slightly compressible, as well as for compressible, flow systems. The technique can be applied to complex and realistic systems such as three-dimensional and multi-phase flow problems.

Even when history matching technique is used to estimate the description data of petroleum reservoirs, the accuracy of the parameter estimates is not guaranteed. There are a number of sources of error in history matching, the major ones being the mathematical model, observability, the criterion function and the match period (2,8). It is, therefore, important to determine the reliability of estimated description data as well as the reliability of both the history and the predicted reservoir performance data.

This paper highlights the reliability aspect of a study made on estimation of reservoir description data by automatic history matching. A method based on non-linear regression theory (1,6) was used to establish the reliability of both the parameter estimates and the predicted performance data, using the estimates. A hypothetical compacting closed square reservoir was studied. Simulated field observed pressures were used for match. In the history matching process, optimal control technique was used to bracket the optimum estimates and then a direct search method was used to narrow the range(s) and to locate the optimum estimates. The mathematical programming method of steepest descent was used in the optimal control segment of the solution. Reservoir parameters such as permeability and porosity were separately and jointly estimated. The confidence limits of the parameter estimates, history performance data and the future performance data generated from the point estimates were established. Joint confidence regions were established for the multi-parameter point estimates in order to account for the joint variabilities of the parameters. The results indicated that parameter estimates by history matching were reasonably reliable. The estimates' reliabilities were indirectly related to the number of unknown parameters. Measurement errors in the matched performance data were found to reduce the accuracy and, hence, the reliability of the estimates.

THEORY

The reservoir simulator developed for this study simulates two-dimensional areal flow of a single-phase, slightly compressible fluid in a closed square reservoir. It is based on the equations developed in previous studies (11,12). The reservoir rock is assumed to be deformable, heterogeneous and anisotropic. The momentum transport equation is given as:

$$\bar{\nabla}[\rho_w \frac{k_{rw}}{\mu_w} k (\bar{\nabla}P - \rho_w g \bar{\nabla}h)] + \frac{\rho_{w_{sc}} q_w}{V_p} =$$

$$\{\phi S_w (\frac{\partial \rho_w}{\partial P})_T + (\rho_w S_w)[C_m + C_{rm}(1-\phi)]\} \frac{\partial P}{\partial t} + (\phi \rho_w) \frac{\partial S_w}{\partial t} + [\phi S_w (\frac{\partial \rho_w}{\partial T})_P] \frac{\partial T}{\partial t} \quad (1)$$

with the applicable initial condition and boundary conditions given in equations (2) and (3), respectively.

$$P(X,Z,t=0) = P_o(X,Z) \quad (2)$$

$$\frac{\partial P}{\partial X} + \rho_w g \frac{\partial h}{\partial X} = 0 \text{ at } X=0 \text{ and } X = LX \quad (3i)$$

$$\frac{\partial P}{\partial Z} + \rho_w g \frac{\partial h}{\partial Z} = 0 \text{ at } Z=0 \text{ and } Z = LZ \quad (3ii)$$

The finite difference method was used to solve the above model equations and the computed well block pressures were converted to bottom hole flowing pressures by using the concept of effective wellbore radius presented by Peaceman (15).

Since optimal control theory was applied in the history matching problem, an adjoint system of equations similar to the state system of equations (equations 1 to 3) was derived. In the adjoint system of equations, Q was substituted for the state primary dependent variable, P. The adjoint system of equations was solved backward in time. The final adjoint equation, the final condition and the associated boundary conditions are shown in equations (4), (5) and (6), respectively.

$$\bar{\nabla}(k\bar{\nabla}Q) - 2\sum_{i=1}^{N} \sum_{j=1}^{M_i} (P^{obs} - P^{cal})\delta(t-t_j) = -\frac{\partial}{\partial t}(\phi' Q) - \frac{\partial}{\partial t}(C' Q) \quad (4)$$

$$Q(r, T_M) = 0 \quad (5)$$

$$\frac{\partial Q}{\partial \gamma}(r_i, t) = 0 \text{ for } r \text{ in } dW_i \quad (6i)$$

$$\frac{\partial Q}{\partial n}(r, t) = 0 \text{ for } r \text{ in } d\omega \quad (6ii)$$

$$\sum_{i=1}^{N} \int_0^{T_M} \int_{dW_i} K' \frac{\partial Q}{\partial n} \delta P_w \, d\delta dt = 0 \quad (6iii)$$

where,

$K' = V_p \frac{\rho_w}{\mu_w} KK_{rw}$

$\phi' = V_p C_w \phi$; $C_w = \frac{\partial \rho_w}{\partial p}$

$C' = V_p \rho_w [C_m + C_{rm}(1-\phi)]$

γ = a point on a well boundary

dW_i = boundary of ith well

w = spartial domain

$d\omega$ = reservoir boundary

The criterion function to be minimized is given as equation (7) while the gradient of the criterion function with respect to reservoir permeability and porosity are given as equations (8i) and (8ii):

$$E = \int_{0}^{T_M} \sum_{i=1}^{n} \sum_{j=1}^{M_i} [P^{obs}(r_i,t) - P^{cal}(r_i,t)]^2 \, \delta(t-t_j)\,dt \tag{7}$$

where $\delta(\xi)$ is the Dirac delta function (Elemo, 1978)

$$\frac{\delta E}{\delta K'} = -\int_{0}^{T_M} (\overline{\nabla} Q - \overline{\nabla} P)\,dt \tag{8i}$$

$$\frac{\delta E}{\delta \phi'} = -\int_{0}^{T_M} P \frac{\partial Q}{\partial t}\,dt \tag{8ii}$$

A necessary condition for optimal solution is that each of the equations (8i) and (8ii) be reduced to zero. As these equations approach zero, the calculated performance data become closer to the history performance data, resulting in optimal parameter point estimates.

The confidence limits equations for parameter estimates and the performance data (history and predicted) are given as equations (9) and (10), respectively.

$$\bar{x}_j - t_{1-\frac{\alpha}{2}} S_{\bar{x}_j} \le \hat{x}_j \le \bar{x}_j + t_{1-\frac{\alpha}{2}} S_{\bar{x}_j} \quad , \; j=1,2,3,\ldots,J \tag{9}$$

$$\bar{P}_{\ell,i_\ell} - t_{1-\frac{\alpha}{2}} S_{\bar{P}_{\ell,i_\ell}} \le \hat{P}_{\ell,i_\ell} \le \bar{P}_{\ell,i_\ell} + t_{1-\frac{\alpha}{2}} S_{\bar{P}_{\ell,i_\ell}} \tag{10}$$

$\ell = 1,2$; $i_1 = 1,2,3 \ldots.$, m; $i_2 = m + 1, m + 2, \ldots, m + I$

where:

J = total number of unknown parameters
m = total number of observation (history) performance data
I = total number of predicted performance data
$\hat{x}_j$ = actual (true) value of jth parameter
$\bar{x}_j$ = estimated (mean) value of jth parameter
$\hat{P}$ = actual (true) value of history and predicted performance data
$\bar{P}$ = model calculated value of history and predicted performance data
S = standard deviation
α = statistical confidence level
t = statistical *t* probability distribution

TABLE I. BASE CASE DATA FOR A HYPOTHETICAL GEOPRESSURED-GEOTHERMAL RESERVOIR

Initial Pressure	=	10,000 psi
Initial Temperature	=	300°F- Isothermal Process
Production Rate	=	40,000 STB/day
Length	=	5,280 feet
Thickness	=	100 feet
Wellbore Radius	=	3 inches
Initial Permeability	=	20 md
Initial Porosity	=	0.20
Density	=	59.401 Lbm/ft^3
Density (Standard Conditions)	=	62.757 Lbm/ft^3
Viscosity	=	0.1993 cp
Formation Volume Factor	=	1.00 RES bbl/STB
Uniaxial Compaction Coefficient	=	9.50×10^{-6} psi^{-1}
Rock Matrix Compressibility	=	0.50×10^{-6} psi^{-1}
Fluid Compressibility	=	3.04×10^{-6} psi^{-1}

TABLE II. PERMEABILITY ESTIMATES AT 0.00% MEASUREMENT ERROR

Type of Estimation	K_ℓ	$\bar{K}$	K_u
K	19.993	19.996	19.999
K & ϕ	16.897	18.998	21.099

TABLE III. PERMEABILITY ESTIMATES AT 0.20% MEASUREMENT ERROR

Type of Estimation	K_ℓ	$\bar{K}$	K_u
K	19.933	19.987	20.042
K & ϕ	18.994	22.828	26.662

A more detailed description of the mathematical model presented herein, the model solution, parameter estimation procedure and the confidence limits equations are described in detail in references 5, 8 and 11.

RESULTS

A set of hypothetical data shown in Table I was assumed true for an undercompacted, stress-sensitive reservoir whose pressures were to be matched. The match period in all runs was 200 days, at which time a pseudo-steady state condition in the reservoir would have been attained. A single producing well was located at the center of the reservoir and was allowed to produce for 200 days. The drawdown data for the 200 days were then matched. Two sets of simulated drawdown data were used. One set was obtained assuming there was no measurement error in the data and another set was obtained with 0.20 percent measurement error. Reservoir permeability and porosity were separately and jointly estimated, and pressure prediction for an additional 60 days was obtained. The confidence limits of the point estimates and the predicted performance (pressure) data at 95 percent confidence level were then calculated. The results are presented as follows:

(a) Reliability of Permeability Estimates

As shown in Table II, the point estimates for permeability are 19.996 md and 18.998 md from single-estimation of permeability and multi-estimation of permeability and porosity, respectively, when error-free drawdown pressure data were matched. The confidence interval from the single estimation is 0.003 md, less than one-tenth percent of the true value of 20 md, implying that the point estimate is very reliable. From the joint estimation, the confidence interval is 2.101 md, about 10.5 percent of the true value, implying much less reliable estimate than the single estimation. Table III shows the permeability point estimates from single and joint estimations when the matched performance data contained 0.20 percent measurement error. The confidence intervals indicate that more reliable estimate was obtained from the single estimation than from the joint estimation, but less reliable estimates were obtained in this case than from the error-free matched performance data case.

Figure 1 shows the confidence limits of the predicted bottom-hole flowing pressures using single-permeability estimate at 0.20 percent measurement error. The confidence interval is about 171.5 psi, less than 3.0 percent of the initial reservoir bottom-hole flowing pressure. The true pressures are all contained within the confidence interval. Figure 2 shows the joint confidence regions for multi-parameter estimation of permeability and porosity at 75, 90, 95 and 99 percent confidence levels and 0.20 percent measurement error. Notice that even at 99 percent confidence level, the confidence interval for permeability is not too wide. Also, the orientation and shape of the ellipses show that there is a strong correlation between permeability and porosity. From all indications, therefore, permeability estimates are reasonably reliable.

(b) Reliability of Porosity Estimates

The point estimates and confidence limits for porosity from error-free simulated history performance data and 0.20 percent measurement error in the simulated history

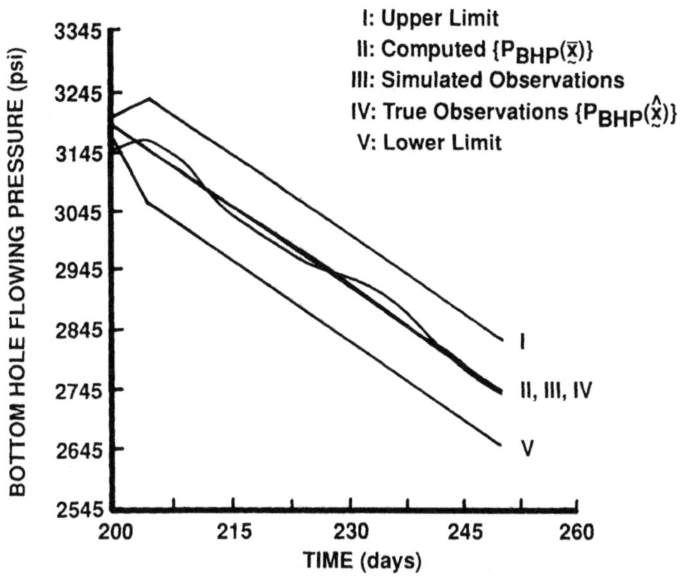

FIGURE 1. CONFIDENCE LIMITS OF PREDICTED PERFORMANCE DATA USING SINGLE-PERMEABILITY ESTIMATES AT 0.20% MEASUREMENT ERROR

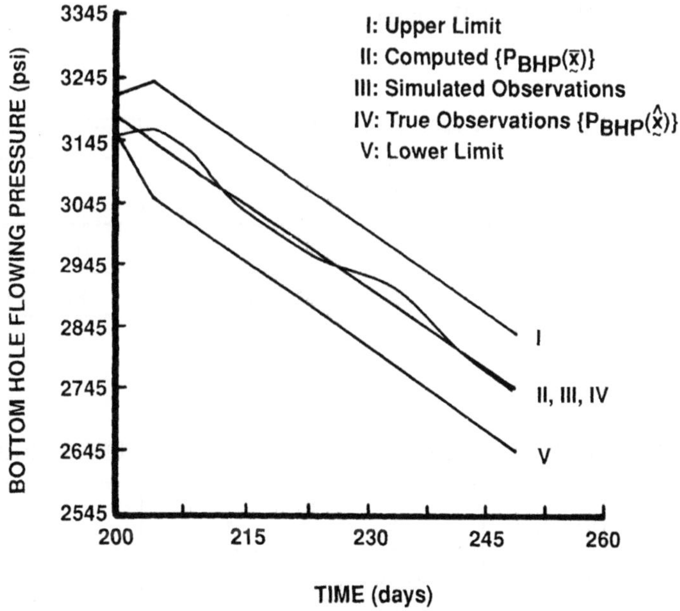

FIGURE 2. CONFIDENCE LIMITS OF PREDICTED PERFORMANCE DATA USING SINGLE-POROSITY ESTIMATES AT 0.20% MEASUREMENT ERROR

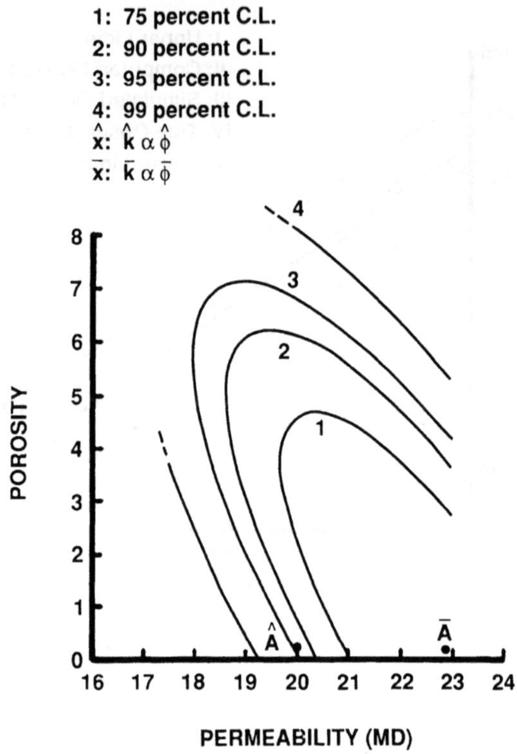

FIGURE 3. JOINT CONFIDENCE REGIONS FOR MULTI-PERMEABILITY AND POROSITY ESTIMATES AT 0.20% MEASUREMENT ERROR

TABLE IV. POROSITY ESTIMATES AT 0.00% MEASUREMENT ERROR

Type of Estimation	ϕ_ℓ	$\bar{\phi}$	ϕ_u
ϕ	0.1975	0.1989	0.2003
ϕ & K	-2.2757	0.1921	2.6599

TABLE V. POROSITY ESTIMATES AT 0.20% MEASUREMENT ERROR

Type of Estimation	ϕ_ℓ	$\bar{\phi}$	ϕ_u
ϕ	0.0655	0.1938	0.3221
ϕ & K	-5.3260	0.1875	5.7010

performance data are shown in Tables IV and V, respectively. In the former case, the porosity point estimate is very good and the confidence interval is very narrow, about 0.70 percent of the true value of 0.20, from single-parameter estimation of porosity. The porosity point estimate from joint estimation of porosity and permeability is reasonably good also; however, the confidence limits, -227.57 and 265.99 percent, are unreasonable. In the latter case, the point estimate from single porosity estimation is also good, but the confidence interval is quite wide, about 64.15 percent of the true porosity value. From joint estimation, the point estimate is 0.1875. Although this is acceptable for practical purposes, it is not as good as the other point estimates. The confidence limits are totally unreasonable.

Figure 3 shows the confidence limits of the predicted bottom-hole flowing pressures using single-porosity estimate at 0.20 percent measurement error. The confidence interval is about 186 psi which is practically acceptable. The true pressures are all contained within the confidence interval also. As can be seen in Figure 2, the confidence regions for joint estimation of porosity and permeability at 0.20 percent measurement error indicate that even at the lowest confidence level of 95 percent, the confidence interval for porosity is very wide. The orientation and shape of the ellipses show that porosity is much less well determined than permeability. It seems, therefore, that porosity estimation is very sensitive to measurement error. Also, porosity estimates are not reliable when joint estimation of porosity and other parameter(s) is made or when there is a significant error in the matched performance data.

DISCUSSION

In most practical simulations, a common major difficulty encountered is the lack of adequate and reliable description data. The conventional well test methods often used to determine such data as permeability do not provide sufficient and accurate description data to overcome this difficulty. History matching technique provides a means of validating and calibrating reservoir simulators, and in the process of calibration, large number of description data can be generated. Consequently, a number of methods have been reported in technical literature for carrying out history matching of different types of reservoirs including undercompacted, stress-sensitive reservoirs. The results of an investigation on the reliability of simulator generated description data are discussed.

The results presented above indicated that some parameters can be reliably estimated for some types of reservoirs by the technique of history matching. To obtain good results from history matching especially when distinguishability and not resolution closure criterion is used, the temporal response of the matched performance data must be sensitive to variations in the description data being estimated. The permeability estimates in this study were found to be accurate and reliable in virtually all cases considered. But porosity estimates were unreliable in most cases. Notice that the reservoir considered in this study is undercompacted and stress-sensitive in which case the reservoir's temporal response of pressure, the matched performance data, is very sensitive to variations in the reservoir permeability while it is only mildly sensitive to variations in the reservoir porosity. For normally compacted and non stress-sensitive reservoirs, the temporal response of pressure is very sensitive to variations in both the reservoir permeability and porosity. It is, therefore, expected that both permeability and porosity can be reliably estimated from the reservoir simulators of those types of reservoirs.

There is a high level of interdependence between reservoir permeability and porosity. This seems to affect porosity estimates more adversely than it affects permeability estimates. Measurement error in matched performance data seems to affect porosity estimates much more adversely also. Therefore, a priori information on both the description and performance data of reservoirs are desirable to improve reliability of the description data estimates, and hence, improve reliability of the predicted performance data. For example, a priori information can be used to determine the error level of history performance data, thereby knowing whether the history data are suitable for matching or not.

CONCLUSIONS

The following conclusions can be drawn based on the foregoing results and discussion:

1. The technique of history matching, based on optimal control theory, can be applied to estimate some of the description data of undercompacted, stress-sensitive reservoirs with greater degree of success than the conventional well test methods.
2. The reliability of reservoir description data estimates, and hence, the predicted performance data, depends on the type of reservoir, the number of unknown description data being estimated, and the type and accuracy of the history performance data to be matched.
3. For an undercompacted, stress-sensitive reservoir, the permeability of the reservoir can be reliably estimated even when there are more than one unknown parameters and there is a little measurement error in the history performance data; but the porosity estimates of the reservoir are not reliable.
4. For a normally compacted, non stress-sensitive reservoir, both the reservoir permeability and porosity can be reliably estimated when error-free history performance data are matched.
5. The reliability of description data estimates is indirectly related to the number of unknown description data being estimated and the level of measurement error in the history performance data being matched.
6. Porosity estimates are most sensitive to measurement error in the history performance data.

NOMENCLATURE

A = Drainage area miles^2
C.L. = Confidence level, percent
C_m = Uniaxial compaction coefficient, psi^{-1}
C_{rm} = Rock matrix compressibility, psi^{-1}

c' = Pseudo-compressibility, $Lbm/_{psi}$
E = Criterion function, psi-psi-days
ESTAR = Minimum E, psi-psi-days
h = Reservoir depth, ft.
K = Absolute permeability, md
K_o = Initial permeability, md
K_{rw} = Relative permeability to water
K' = Pseudo-permeability, $\frac{Lbm - md}{cp}$
m_i = Number of observation data from i^{th} well
n = Total number of wells in the reservoir
P = Pore pressure, psi
p^{obs} = Observed pressure data, psi
p^{cal} = Calculated pressure data, psi
P_o = Initial reservoir pressure, psi
q = Flow rate, STB/day
Q = Adjoint variable, psi/ft
S_w = Water saturation
t = Time, days
T = Reservoir temperature, degree F.
T_M = Match period, days
v_p = Block bulk volume, ft^3
w = Reservoir net pay thickness, ft
w_o = Initial reservoir net pay thickness, ft
ϵ = Performance data measurement error
ϕ = Porosity, fraction
ϕ_o = Initial porosity, fraction
ϕ' = Pseudo-porosity, Lbm/psi
ρ_w = Water density, Lbm/ft^3
ρ_{sc} = Water density at standard conditions, Lbm/ft^3
μ_w = Water viscosity, cp
$\hat{x}$ = True (actual) value of parameter x
$\bar{x}$ = Estimated (mean) value of parameter x
x_l = Lower confidence limit of parameter x estimate
x_u = Upper confidence limit of parameter x estimate

REFERENCES

1. Bard, Y., *Nonlinear Parameter Estimation*, Academic Press, New York, NY, 1974.
2. Chavent, C., M. Dupuy and P. Lemmonnier, "History Matching by Use of Optimal Control Theory," *Society of Petroleum Engineers Journal*, February, 1975, pp. 74-86; *Trans*. AIME, Vol. 259.
3. Coats, K.H., "Use and Misuse of Reservoir Simulation Model," *Journal of Petroleum Technology*, November 1969, pp. 1391-1398.
4. Coats, K.H., "Reservoir Simulation: State-of-the-art," *Journal of Petroleum Technology*, August, 1982, pp. 1633-1642.

5. Dogru, A.H., *Confidence Limits on the Parameters and Predictions of One Dimensional Single Phase Slightly Compressible Reservoirs*, Ph.D. Dissertation, The University of Texas at Austin, May 1974.

6. Draper, N.R. and H. Smith, *Applied Regression Analysis*, John Wiley and Sons, Inc., New York, New York, 1966.

7. Earlougher, Jr., R.C., *Advances in Well Test Analysis*, Society of Petroleum Engineers of AIME Monograph, Vol. 5, 1977.

8. Elemo, R.O., *Automatic History Matching of Geopressured-Geothermal Reservoirs Performance Data by Use of Optimal Control Theory*, Ph.D. Dissertation, The University of Texas at Austin, December 1978.

9. Elemo, R.O. and R.M. Knapp, "Estimation of Stress-Sensitive Reservoirs' Initial Parameters by Use of the Conventional Well Test Methods," Paper presented at the Fourth United States Gulf Coast Geopressured-Geothermal Energy Conference: Research and Development, held at The University of Texas at Austin, Texas, October 29-31, 1979.

10. Elemo, R.O. and R.M. Knapp, "Numerically Controlled Experiments on Initial Parameters for Stress-Sensitive Reservoirs," *Revista Técnica*, 1981, Vol. 4, No.1.

11. Elemo, R.O. and R.M. Knapp, "Estimation of Stress-Sensitive Reservoirs' Initial Parameters by Automatic History Matching," *Revista Técnica*, 1982, Vol. 5, No. 1.

12. Knapp, R.M., O.F. Isokrari, S.K. Garg and J.W. Pritchett, "An Analysis of Production from Geopressured Aquifers," SPE Paper No. 6825, presented at the SPE-AIME Fall Meeting, held at Denver, Colorado, Oct. 9-12, 1977.

13. Lions, J.L., *Optimal Control of Systems Governed by Partial Differential Equations*, Translated into English by S.K. Mitter, Springer-Verlag, New York, New York, 1971.

14. Odeh, A.S., "Reservoir Simulation....What is it?," *Journal of Petroleum Technology*, November 1969, pp. 1383-1388.

15. Peaceman, D.W., "Interpretation of Well-Block Pressures in Numerical Reservoir Simulation," *Society of Petroleum Engineers Journal*, June 1978, pp. 183-194; *Trans*. AIME, 253.

16. Saleri, N.G. and R.M. Toronyi, "Engineering Control in Reservoir Simulation: Part I," SPE Paper No. 18305, presented at the Society of Petroleum Engineers Fall Meeting, held at Houston, Texas, October 2-5, 1988.

17. Watson, A.T., J.H. Seinfeld, G.R. Gavalas and P.T. Woo, "History Matching in Two-Phase Petroleum Reservoirs," *Society of Petroleum Engineers Journal*, December 1980, pp. 521-532.

ABRASION EMPIRICAL EQUATIONS FOR THE ITABIRITE MUD SYSTEM

Aboulghasem Moonesan
Mohamed S. Bizanti

Petroleum Engineering Department
Louisiana Tech University
Ruston, LA

Twenty-eight mud samples were tested in the laboratory to predict abrasion equations that will aid mud engineers for better treatment of itabirite mud systems. Three different ranges of particle size of itabirite were selected for the tests. Particle sizes between 74 and 125 microns, between 44 and 74 microns, and between 37 and 44 microns were used. Thirty-minute abrasion tests were conducted with a modified Hamilton Beach Blender at 9800, 11200, 12750, and 14350 RPM. Rheological properties, filtration properties, and abrasiveness were studied and analyzed. Regression analyses were performed between abrasion and the parameters of size, polymer concentration (Drispac Superlo or Kelzan XCD), and shear rate investigated in this research.

Empirical relations relating the abrasion to the shear rate, particle sizes and polymer concentrations were generated using linear, semi-log, and log-log models. The log-log model appeared to best fit the data obtained. The Empirical equations can be utilized to predict abrasion associated with other mud systems.

INTRODUCTION

Since the 1940's, barite has been the most dominant material for weighting muds. The reasons are its fairly high specific gravity, low abrasiveness, usefulness as inert material in both water-base and oil-base muds, and the availability at a low cost.

In the United States over 90% of the consumption of barite is a direct result of the need for weighted muds in oil and gas wells drilled. Therefore, barite consumption depends on the price of oil and gas and on drilling activity. Recently, with the unprecedented drop in the price of oil and gas, the drilling decreased, as did the demand for barite. Based on the assumption that oil prices will rise again, the shortage of high-quality barite will cause a significant increase in the cost of

Particle Technology and Surface Phenomena in Minerals and Petroleum
Edited by M.K. Sharma and G.D. Sharma, Plenum Press, New York, 1991

drilling fluid. As a result, the petroleum industries will face the same problems they experienced in 1981. The estimated consumption of barite in the drilling industry was 3 million tons in 1981, a significant increase from the 1980 figure of 2.50 million tons. Barite demand in the petroleum industry is projected to increase from 2.65 million tons in 1983 to 5.0 million tons in the year 2000. The United States reserves of barite, estimated at 30 million tons, are not adequate to fill the projected cumulative demand of about 60 million tons by the year 2000.[1,2]

Barite is not readily available near centers of petroleum activities. The United States imports barite from South American countries as well as from Asian countries such as India and China. By 1983, high-quality Chinese barite was fulfilling nearly 60% of the U.S. import requirements. A major problem is the transportation of this heavy mineral for long distances, which significantly increases the price.[1]

Screen Analysis

- Minus 100 mesh	100%
- Minus 200 mesh	98.5% - 99.8%
- Minus 325 mesh	92.0% - 96.0%
- Free moisture @ 105°	Less than 0.15%
- Color	Brown to red brown
- Specific gravity	5.1
- Magnetite Content	Trace
- pH	6.5 - 7.5

Typical Chemical Analysis

- Fe	68.0% - 68.5% (92.2% - 97.9% Fe_2O_3)
- SiO_2	0.5% - 1.0%
- Al_2O_3	0.5% - 1.0%
- CaO + MgO	0.15%
- S	0.01%
- P	0.03%
- Cu	0.01%
- Na_2O	0.02%
- K_2O	0.005%
- Ca^{++}	10.0 - 15.0 PPM
- Mg^{++}	0.5 - 2.0 PPM

Barite is frequently contaminated with alkaline-soluble carbonate and sulfide minerals that cause serious drilling fluid problems. Therefore, small concentrations of impurities in barite can lead to significant contamination of the drilling fluid.[3]

Drispac Superlo is a chain polymer, shorter than regular Drispac, which is a polyionic cellulose derivative. This polymer is manufactured and marketed by Drilling specialties Company (1985). Drispac Superlo is dispersible in water-base mud and functions in both fresh and salt water muds.

Mud Formation

Twenty-eight mud types were investigated during this research (Table 1). All the mud types contained the same amount of 14 lbm/bbl bentonite. It was determined that in the itabirite mud system 16% and in the barite mud system 20.5% inert solids by total volume were required to obtain the desired mud density of 14ppg. The difference in the solid percent in these two muds was mainly due to the difference in specific gravities.

Except for mud types B, IL, IM, and IS (untreated muds, Table 1), 0.5 lbm/bbl (0.5 g/350 ml) Desco thinner was used to (a) minimized the effect of viscosity, where a selected polymer (Drispac Superlo, or Kelzan XCD) was used in concentrations of 0.25, 0.5, and 0175 lbm/bbl as a coating agent to lower the mud abrasiveness, and (b) to reduce the yield point, as Desco thinner has the tendency to decrease the mutual attraction of the active clay particles through absorption on the clay surfaces. A 0.25 lbm/bbl caustic soda was used in all treated muds for pH control

Chilingarian and Vorabut (1983) have stated athat the pH of water-base mud is controlled by the addition of caustic soda (NaOH). In general, a pH value below 8 and above 11 promotes floccuation that causes increases in fluid loss, viscosity, and gel strength. A pH value ranging between 8 and 11 tends to promote dispersion. The amount of caustic sode used in all treated muds was 0.25 lbm/bbl.

Test Equipment Used

Standard mud laboratory equipment was utilized in this research. Mud density was determined by a Baroid mud balance. Mud rheology was determined by a Bariod variable-speed electronic rheometer and a standard filter press. Presently, there is no API approved method for determining abrasiveness of a

Table 1. Drilling Mud Types Tested

Mud Type	Mud Composition	Size Distribution
B	(1.54% Bent + 20.5% Bar + 77.96% Dis W)	API
B1	(Type B) + .5 ppb D + .25 ppb NaOH + .25 ppb DS Pol	API
B2	(Type B) + .5 ppb D + .25 ppb NaOH + .5 ppb DS Pol	API
B3	(Type B) + .5 ppb D + .25 ppb NaOH + .75 ppb DS Pol	API
B4	(Type B) + .5 ppb D + .25 ppb NaOH + .25 ppb KXCD Pol	API
B5	(Type B) + .5 ppb D + .25 ppb NaOH + .5 ppb KXCD Pol	API
B6	(Type B) + .5 ppb D + .25 ppb NaOH + .75 ppb KXCD Pol	API
IL	(1.54% Bent + 16.0% Itab + 82.46% Dis W)	74-125 microns
IL1	(Type IL) + .5 ppb D + .25 ppb NaOH + .25 ppb DS Pol	74-125 microns
IL2	(Type IL) + .5 ppb D + .25 ppb NaOH + .5 ppb DS Pol	74-125 microns
IL3	(Type IL) + .5 ppb D + .25 ppb NaOH + .75 ppb DS Pol	74-125 microns
IL4	(Type IL) + .5 ppb D + .25 ppb NaOH + .25 ppb KXCD Pol	74-125 microns
IL5	(Type IL) + .5 ppb D + .25 ppb NaOH + .5 ppb KXCD Pol	74-125 microns
IL6	(Type IL) + .5 ppb D + .25 ppb NaOH + .75 ppb KXCD Pol	74-125 microns
IM	(1.54% Bent + 16.0% Itab + 82.46% Dis W)	44-74 microns
IM1	(Type IM) + .5 ppb D + .25 ppb NaOH + .25 ppb DS Pol	44-74 microns
IM2	(Type IM) + .5 ppb D + .25 ppb NaOH + .5 ppb DS Pol	44-74 microns
IM3	(Type IM) + .5 ppb D + .25 ppb NaOH + .75 ppb DS Pol	44-74 microns
IM4	(Type IM) + .5 ppb D + .25 ppb NaOH + .25 ppb KXCD Pol	44-74 microns
IM5	(Type IM) + .5 ppb D + .25 ppb NaOH + .5 ppb KXCD Pol	44-74 microns
IM6	(Type IM) + .5 ppb D + .25 ppb NaOH + .75 ppb KXCD Pol	44-74 microns
IS	(1.54% Bent + 16.0% Itab + 82.46% Dis W)	37-44 microns
IS1	(Type IS) + .5 ppb D + .25 ppb NaOH + .25 ppb DS Pol	37-44 microns
IS2	(Type IS) + .5 ppb D + .25 ppb NaOH + .5 ppb DS Pol	37-44 microns
IS3	(Type IS) + .5 ppb D + .25 ppb NaOH + .75 ppb DS Pol	37-44 microns
IS4	(Type IS) + .5 ppb D + .25 ppb NaOH + .25 ppb KXCD Pol	37-44 microns
IS5	(Type IS) + .5 ppb D + .25 ppb NaOH + .5 ppb KXCD Pol	37-44 microns
IS6	(Type IS) + .5 ppb D + .25 ppb NaOH + .75 ppb KXCD Pol	37-44 microns

Table 2. Properties of Mud Types (Mud Weight = 14 ppg)

Mud Type	PV[1] (cp)	YP[1] (1bf/100 sq ft)	GELS[1] (1bf/100 sq ft)	Corrected Cumu.[2] Fluid Loss (ml)	pH Value
B	23	7	12/25	17.6	7.8
B1	25	9	8/12	15.2	10..6
B2	29	10	7/10	10.2	10.8
B3	34	12	5/6	7.9	10.7
B4	33	12	11/14	17.3	10.8
B5	42	22	8/10	11.5	10.9
B6	50	31	5/7	9.4	10.9
IL	16	3	9/15	23.4	8.4
IL1	20	4	7/11	18.0	10.8
IL2	24	6	6/9	15.5	10.7
IL3	29	7	3/5	13.8	10.7
IL4	24	7	8/12	20.3	10.9
IL5	31	14	7/10	15.4	10.8
IL6	37	22	5/8	14.7	10.7
IM	19	5	10/16	25.9	8.4
IM1	22	7	8/12	22.7	10.8
IM2	26	6	6/11	19.6	10.7
IM3	31	9	4/7	15.8	10.8
IM4	27	7	8/14	22.8	10.8
IM5	32	17	7/12	21.8	10.9
IM6	38	26	6/9	17.5	10.8
IS	20	6	12/18	27.6	8.47
IS1	23	8	8/14	24.9	10.9
IS2	27	8	7/12	22.1	10.8
IS3	32	11	5/8	16.2	10.8
IS4	29	11	7/14	26.1	10.7
IS5	33	19	6/12	24.7	10.7
IS6	39	28	6/10	19.1	10.8

[1] measured at 74°F
[2] measured at 74°F and 100 psi

fluid of oil field material. A modified Hamilton Beach 14-speed blander was used to conduct the abrasion tests. Other auxiliary equipment such as a sieve shaker and Beckman zeromatic pH meter were also utilized.

Abrasion Tests

Initial weight of the blade to be tested was weighted to ten-thousandths of a gram. After loading the binder with 500 cc of test mud sample, it was stirred for 30 minutes at various readings of 9800, 11200, 12750, and 14350 RPM. At the end of each 30 minute period, the blade was removed, cleaned, dried and reweighed to ten-thousandths of a gram. The weight loss of the blade in grams and percent weight loss at each RPM was determined.

INTERPRETATION AND DISCUSSION OF EXPERIMENTAL RESULTS

Rheological Results

The results of rheological and filtration properties are presented in Table 2 for all mud types. It is apparent from the results that a treated or untreated itabirite mud systems has better rheological characteristics than a barite mud system (untreated or treated). This is due to the fewer inert solid particles in the itabirite mud system that cause less mechanical interaction and electro-chemical attraction between the active particles. With the addition of polymer at each concentration, lower PV and YP can be obtained with the itabirite mud system than with barite. Therefore, the use of polymers in the itabirite mud system were more compatible than in the barite mud. This advantage would enable the itabirite mud system to be tested with an even higher polymer concentration, overcoming the abrasion problem in the mud system.

Itabirite mud system exhibited higher cumulative fluid loss volumes than barite mud system. It was observed that 0..5 lbm/bbl concentration of DS or KXCD polymer for a large particle size of itabirite (between 74 and 125 microns), and 0.75 lbm/bbl for small particle size of itabirite (between 34-44 microns) were adequate to improve the filtration properties.

Abrasion Results

Abrasion was measured as weight loss (in grams) in the blinder's test blades. It was assumed that the greater the weight loss, the greater the abrasiveness of the mud. The percent weight losses were calculated for all mud types and are presented for each mud type as selected shear rates in Table 3.

It can be seen from Table 3 that abrasion increased in all mud type when the rotational speed of the blade was increased. It is interesting to note that in some cases at higher velocity (14,350 RPM), smaller ranges of particle size of itabirite were slightly less abrasive than barite at each polymer concentration. Barite crystals have a greater number of sharp edges than itabirite.[2] Hence, abrasivity of the barite was probably due to the shape and nature of the mineral particles. In addition, in a mud system weighted with itabirite, a lower solid content can be used to obtain the proper weight. Thus, a polymer can be more effective as a coating agent between the interacting surfaces.

Resulting weight losses of mud types B, IL, IM, and IS with no polymer concentration (untreated mud, Table 3) indicated that the selected ranges of particle size of itabirite did not lower the abrasiveness to the barite level, but a more acceptable level of abrasion appeared. Itabirite particle sizes between 37 and 44 microns showed more than 55% lower abrasion compared to the particle sizes between 74 and 125 microns. In addition, ranges of particle size of itabirite between 44 and 74 microns showed more than 43% lower abrasion compared to particle sizes between 74 and 125 microns. Abrasiveness was seen to have a similar dependence on the particle size distribution in a mud system. The smaller the solid particle sizes contained within the mud of the same density, the less abrasive it was.

In this study, Drispac Superlo or Kelzan XCD polymer with concentration of 0.25, 0.5, and 0.75 lbm/bbl were used as coating agents for the purpose of lowering the abrasiveness. It is apparent from the results (Table 3) that 0.75 lbm/bbl concentration of DS or KXCD for large particle sizes (74-125 microns) and 0.5 lbm/bbl for small particle sizes (37-44 microns) were adequate for lowering the abrasivity of the itabirite mud system to the barite level.

Regression Analysis

It was postulated in this study that abrasion in a mud system could be demonstrated as a function of mud density, shear rate (RPM), polymer concentration (PC, lbm/bbl), stirring time (min), and particle size (s, millimeters) associated with the mud system. In this study, two of the independent variables (mud density and stirring time) were held constant.

Regression analysis techniques were applied to determine the regression coefficients associated with abrasion as a function of particle size, polymer concentration, and shear rate in the following three models:

Table 3. Abrasion Tests (%) Wt. Loss of Mud Types at Selected Shear Rates

Mud Type	9800 (RPM)	11200 (RPM)	12750 (RPM)	14350 (RPM)
B	0.11	0.18	0.25	0.34
B1	0.09	0.16	0.24	0.30
B2	0.06	0.14	0.20	0.27
B3	0.05	0.09	0.15	0.19
B4	0.09	0.15	0.22	0.29
B5	0.07	0.13	0.19	0.23
B6	0.04	0.08	0.13	0.16
IL	0.33	0.33	0.72	10.7
IL1	0.28	0.32	0.37	0.46
IL2	0.21	0.25	0.31	0.44
IL3	0.18	0.22	0.28	0.34
IL4	0.21	0.25	0.31	0.45
IL5	0.19	0.22	0.28	0.31
IL6	0.16	0.18	0.23	0.26
IM	0.25	0.28	0.37	0.49
IM1	0.23	0.26	0.31	0.37
IM2	0.20	0.24	0.28	0.33
IM3	0.13	0.17	0.21	0.24
IM4	0.19	0.22	0.26	0.32
IM5	0.16	0.21	0.23	0.27
IM6	0.15	0.18	0.20	0.22
IS	0.19	0.25	0.31	0.36
IS1	0.17	0.21	0.23	0.25
IS2	0.14	0.16	0.19	0.21
IS3	0.13	0.15	0.18	0.19
IS4	0.15	0.18	0.25	0.29
IS5	0.12	0.14	0.17	0.19
IS6	0.10	0.12	0.15	0.17

Table 4. Prediction of Abrasion Equations

Polymer	Linear Model	R^2
DS	$AB = -0.32304979 + 2.94808731(S) - 0.27266667(PC) + 0.000043049(RPM)$	0.6728
KXCD	$AB = -0.28335346 + 2.50394665(S) - 0.30566667(PC) + 0.000041234(RPM)$	0.6345
DS &KXCD	$AB = -0.21134884 + 2.26463936(S) - 0.26089744(PC) + 0.000035677(RPM)$	0.6523
Polymer	**Semi-log Model**	**R^2**
DS	$AB = 0.04059807\ {}_e8.80979373(S)\ {}_e\text{-}0.83701060(PC)\ {}_30.000132340(RPM)$	0.9027
KXCD	$AB = 0.03957435\ {}_e7.82901398(S)\ {}_3\text{-}1.00257323(PC)\ {}_30.000136564(RPM)$	0.8895
DS &KXCD	$AB = 0.04454386\ {}_e7.78058454(S)\ {}_e\text{-}0.87516734(PC)\ {}_e0.000126484(RPM)$	0.8735
Polymer	**Log-log Mode**	**R^2**
DS	$AB = 0.00000249(S)^{0.53060128}(PC)^{-0.31992027}(RPM)^{1.35026810}$	0.9350
KXCD	$AB = 0.00000091(S)^{0.44662252}(PC)^{-0.32860506}(RPM)^{1.41764398}$	0.9340
DS &KXCD	$AB = 0.00000150(S)^{0.18861190}(PC)^{-0.32426267}(RPM)^{1.38395604}$	0.8906

$$LINEAR\ MODEL: \quad AB = b_o + b_1(S) + B_2(PC) + b_3(RPM) \tag{1}$$

$$SEMI-LOG\ MODEL: \quad AB = b_o e^{b_1(s)} * e^{b_2(PC)} * e^{b_3(RPM)} \tag{2}$$

$$LOG-LOG\ MODEL: \quad AB = b_o (S)^{b_1} (PC)^{b_2} (RPM)^{b_3} \tag{3}$$

Regression analysis results are summarized in Table 4. The regression analysis show that there is a good correlation between the dependent variable (AB) and the independent variables for all investigated models in this research. The log-log model is believed to be the best suited for describing the laboratory data due to high value of R^2 (coefficient of multiple determination). Results show that shear rate is the most influential parameter on abrasion. Particle size was found to have a moderate effect on abrasion. The log-log regression model also indicates that an increase in PC lowers abrasion. The regression coefficient associated with PC in the abrasion equations indicates that the abrasion is less sensitive to PC as compared to S and RPM.

NOMENCLATURE

AB	=	Abrasion
b_0, b_1, b_2, b_3	=	The regression coefficients of the model that were determined by applying regression analysis
B	=	Bar = Barite
Bent	=	Bentonite
cp	=	Centipoise
Cumu	=	Cumulative
Dis W	=	Distilled Water
D	=	Desco
DS	=	Drispac Superlo
GELS	=	Gel Strength
IL	=	Itabirite Large Particle Sizes (74 -125 microns)
IM	=	Itabirite Medium Particle Sizes (44 - 74 microns)
IS	=	Irabirite Small Particles Sizes (37 - 44 microns)
Itab	=	Itabirite
KXCD	=	Kelzan XCD
PC	=	Polymer Concentration
Pol	=	Polymer
ppb	=	Pounds Per Barrel
ppg	=	Pounds Per Gallon

PV	=	Plastic Viscosity
RPM	=	Revolutions Per Minute
S	=	Particle Size
Sec	=	Seconds
YP	=	Yield Point

REFERENCES

1. Sarkis, G. Ampian, *"Barite,"* Mineral Facts and Problems, Bureau of Mines, Bulletin 675, pp. 56-73, 1985.
2. Bokle, F. M., *"Techniques for Evaluation and Improving the performance of the Iron Oxides Used as Barite Substitutes in Oil Well Drilling Fluids,"* Ph.D. Dissertation, University of Oklahoma, 1982.
3. Binder, G. G. Jr., et al., *"Evaluating Barite as a Source of Soluble Carbonate and sulfide Contamination in Drilling Fluids,"* Journal of Petroleum Technology, Dec., pp. 2371-2375, 1981.
4. Scharf, A. D., and Watts, R. D., 'Itabirite: an Alternative Weighting Material for Heavy Oil-Base Muds," Paper SPE 13159, presented at the 1984 SPE Annual Technical Convention and Exhibition Houston, Sept., 1984.
5. Golis, S. W., *"Oil Mud Techniques Improve Performance in Deep,"* presented at the 1984 SPE Annual Technical Conference and Exhibition, Houston, TX, Sept. 17-21, 1984.
6. Walker, C. O., *"Alternative Weighting Material"* Journal Pet. Tech. pp. 2158-2164, Dec., 1983.
7. Demsimix Technical Data, 1313 First City National Bank Building, Houston, TX 77002, June, 1983.
8. Product Literature, *"Drispac Superlo Additive", Desco,* Drilling Specialties Company, Technical Services Division, Bartlesville, OK 74004, 7 pp. 1985.
9. Kelco Rotary Drilling Fluid Division, Kelco Oil Field Group, Inc. KR-1-A, Copyright Kelco Oil Field Group, Inc., 1982.

CORRECTING OIL-WATER RELATIVE PERMEABILITY DATA FOR CAPILLARY END EFFECT IN DISPLACEMENT EXPERIMENTS

S. Qadeer, K. Dehghani, D. O. Ogbe and R. D. Ostermann

Petroleum Development Laboratory
University of Alaska Fairbanks
Fairbanks, AK

Capillary effects in low flow rate displacement experiments cause error in relative permeability values calculated by JBN or Jones and Roszelle methods. The results from a series of displacement and capillary pressure experiments were used along with a history matching package to quantify the changes in relative permeabilities with rate. The experiments were conducted with different rates on short and long cores. The history matching procedure was developed using a one-dimensional, two phase finite difference simulator, and the Levenberg-Marquardt optimization algorithm.

Exponential form for relative permeability functions were used and the change of function parameters with rate was investigated. The results show that for the case of drainage the relative permeability parameters (i.e., end-point relative permeability and saturation exponent) have functional relationships with rate. The saturation exponent in non-wetting phase and the end-point non-wetting relative permeability increase with rate. However, the end-point relative permeability to wetting phase did not show any significant change but the saturation exponent in the wetting phase decreased with increase in rate. Regression equations of these changes were developed. In the case of imbibition, these parameters did not indicate any meaningful trend with rate.

INTRODUCTION

Relative permeabilities are measured in the laboratory by two general methods, steady state and unsteady state displacement experiments. Although steady state experiments allow data collection over the entire saturation range, the fact that fluid distribution in the steady state experiments may not be the representative of a displacement process, makes the unsteady state experiments more applicable to the reservoir case, (Heaviside and Black (1983)).

Particle Technology and Surface Phenomena in Minerals and Petroleum
Edited by M.K. Sharma and G.D. Sharma, Plenum Press, New York, 1991

The recovery and pressure data from displacement experiments are used to calculate the relative permeabilities by Johnson, Bossler and Naumann (JBN) method (1959) or by a graphically equivalent method of Jones and Roszelle (1978). These two methods are based on the assumption that capillary forces in the displacement process are negligible. The existence of capillary forces affects the recovery and pressure data and causes erroneous results from the above methods.

In displacement experiments, it is the practice to minimize the capillary end effect by setting the scaling coefficient, $L\upsilon\mu_w$, above the range of 1-5 Cm^2/min-cp (after Rapopart and Leas (1959), and Kyte and Rapopart (1958)). This means that displacement experiments have to be conducted with rates much higher than reservoir rates.

Englebert and Klinkenberg (1951), Chouke et al (1959), and Peters and Flock (1981), show that higher rates in the laboratory displacement experiments can cause instability of the front. Fines migration has been reported by Saraf and McCaffery (1981) and Gabriel and Inamdar (1983) to be another source of problem at the higher rates. In addition to these problems, the relative permeability values calculated by JBN method in strongly water wet systems tend to be inaccurate because of piston like displacement, (Saraf and McCaffery (1981)) and Archer and Wang (1973).

To overcome the above problems, it is preferred that the displacement experiment be conducted with reservoir rates and relative permeabilities can be found by taking into account the effect of capillarity on recovery and pressure data. Sigmund and McCaffery (1979) and a later extension of their work, Batycky et al (1981) have developed a method to characterize relative permeability curves by two parameters. They used these relative permeabilities in a simulator which included the effect of capillary pressure to calculate pressure drops and recoveries. Using a Newton Raphson iteration procedure, they varied the parameters until the least square error between the measured and the computed pressure drops and recoveries from the simulator was minimum. Using the parameter estimation approach Kerig and Watson (1985) represented the relative permeabilities by cubic spline functions and subjected these functions to inequality constraints. They used the Marquardt's modification of Gaus Newton minimization algorithm to estimate the parameters. Tao and Watson (1984) have studied the effect of different experimental operating conditions (i.e. random fluctuations in the injection rate, pressure drop, and oil production measurement) on the accuracy of JBN relative permeability values. They also showed that when a computer algorithm is developed for JBN the type of function selected to represent the production data could introduce error in differentiation. Odeh and Dotson (1985) presented a method for correction of relative permeability values calculated by Jones and Roszelle from low rate displacement data. Their method of correction was based on the postulation that q_o/K_{ro} or q_w/K_{rw} and average water saturation normally have a linear relationship at higher rates and lower saturation ranges. Civan and Donaldson (1987) presented a method to calculate relative permeabilities. Their method is based on simultaneous solution of the of the fractional flow and an integro-differential equation developed from the material and momentum balance equations for two phase flow including capillary pressure.

The objective of this study is to determine if a functional relationship exists between rate and error caused by capillary forces on relative permeabilities at different saturations. Only strongly water wet system is considered. To accomplish this objective, the following steps were taken:

1. A series of displacement experiments were conducted with different rates on long and short cores. Also, steady State experiments were conducted for comparison.

2. A history matching package consisting of a one-dimensional, two-phase, finite difference simulator and Levenberg-Marquardt optimization algorithm was developed. The capillary end effect is considered in the simulator.

3. Experimentally determined capillary pressure and displacement production data were used to find relative permeability functions which we call "true" relative permeabilities.

4. Using the correct relative permeability functions and the simulator, recovery and pressure data for different displacement rates were calculated.

5. The calculated production data were used to get relative permeability functions using the optimization algorithm and a non-capillary welge type model. We call these relative permeabilities as "false", and we compare them with true relative permeabilities for a possible quantification of how the error varies with rate.

Laboratory Experiments

The experimental set up is shown on Figure 1. The fluids were pumped by two dual cylinder positive displacement pumps (FDS-210, Petrophysical Services)

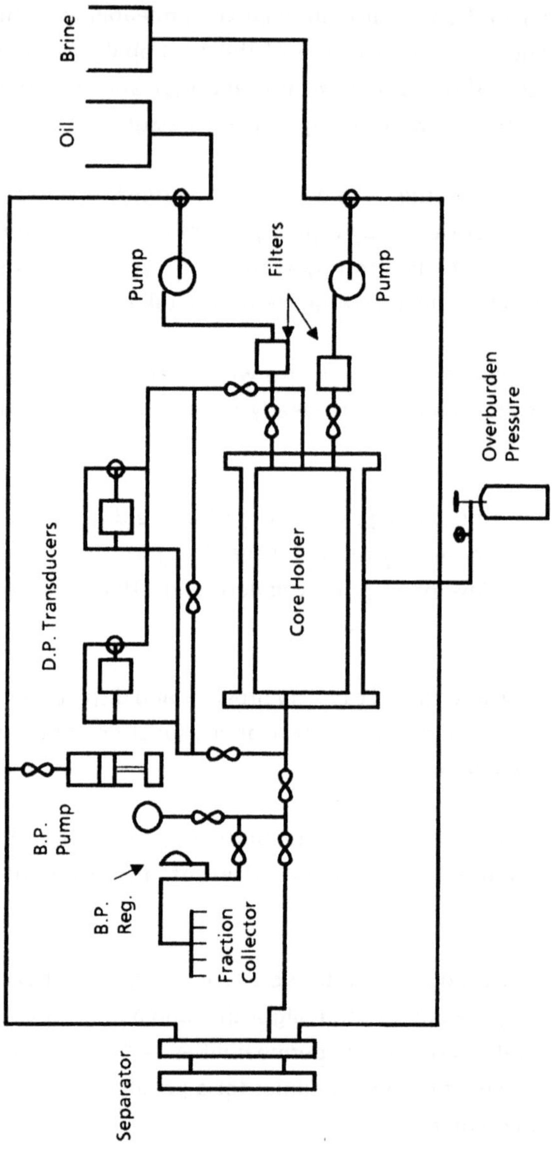

Fig. 1. **Experimental setup.**

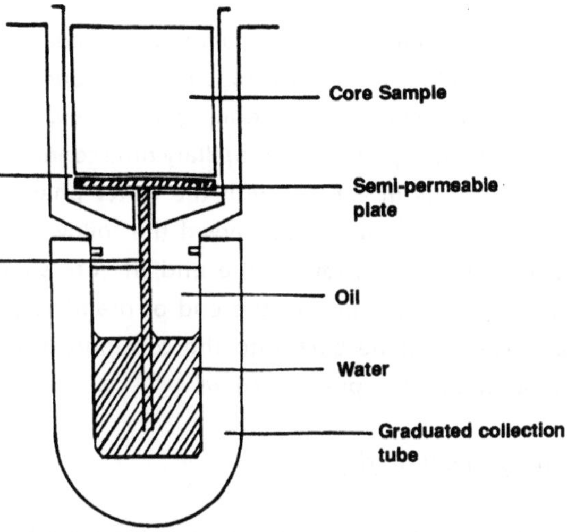

Fig. 2. Modified bucket for drainage and imbibition capillary pressure measurements.

Table 1. Properties of the Core Samples

	Length (cm)	Absolute Permeability (md)	Porosity %
Core No. 1	24.1	528	24.4
Core No. 2	25.4	980	22.8
Core No. 3	12.7	950	23.0

into a core holder. The core holder was placed in the air bath for a constant temperature (25 ± 0.5°C). The production was measured by an acoustic interface monitoring separator (Petrophysical Services). For steady state experiments oil and water were recycled from the separator to core and for the unsteady state displacements, only the displacing fluid was recycled. The produced fluid could be measured with ± 0.1 Cm^3 accuracy and the pressure difference across the core sample was measured by pressure transducers.

The capillary pressure curves were measured with an ultracentrifuge (Beckman L8M/P). The centrifuge drainage buckets were modified so that both drainage and positive imbibition capillary pressure curves could be measured. A semi-permeable plate with a capillary tube connected to it kept the core outflow face constantly in contact with the water phase in the measuring tube (see Figure 2). This modification served the following purposes: first it assured a 100% water saturation at the end of core sample, and second when the centrifugal acceleration at the end of drainage process was reduced in steps, water could imbibe back into the core and this allowed us to measure the imbibition capillary pressure curve.

Material and Procedure

A total of three core samples were used for all the experiments. These were one inch diameter fired Berea Sandstone. The properties of these cores are shown in Table 1. The oil phase was Kerosene and the aqueous phase was 2% brine. Kerosene was depolarized through a silica gel column, and degased before use in the experiments.

The core samples were flooded with CO_2 for several hours and saturated with brine under vacuum. To insure complete saturation and stabilization of absolute permeability each core was flooded with at least 300 pore volumes of brine while a back pressure of 500 psi was always imposed against the flow. For all the cores primary drainage was completed by displacement. Oil was pumped with a rate of 2 cc/min for 150 pore volumes. The water saturation at this stage was considered irreducible. Tables 2 through 4 each show the types of experiments, the order in which they have been conducted, and the end points of each experiment for core number 1 to 3 respectively. The displacement experiments conducted with a rate lower than 2 cc/min, were followed by a 'bump' after the pressure drop and production was stabilized. For these experiments, the end point relative permeabilities and water saturation after the 'bump', and the average water saturation before the 'bump' are shown. Steady state experiments were only conducted on Core No. 1. These experiments

Table 2. Experiments Conducted on Core No. 1

Experiment No.	Experiment Type	Rate cc/min	$S_{w(end)}$ After Bump	$Kr_{w(end)}$	$Kr_{o(end)}$	Average S_w Before Bump
1	Displacement Primary Drainage	2	0.23	---	0.822	---
2	Displacement Imbibition	1	0.67	0.058	---	---
3	Displacement Drainage	1	0.27	---	0.635	0.32
4	Displacement Imbibition	1	0.66	0.053	---	0.65
5	Displacement Drainage	0.5	0.28	---	0.661	0.37
6	Displacement Imbibition	0.5	0.65	0.046	---	0.64
7	Steady State Drainage	2	0.28	---	0.69	---
8	Steady State Imbibition	2	0.69	0.04	---	---

Table 3. Displacement Experiments Conducted on Core No. 2

Experiment No.	Experiment Type	Rate cc/min	$S_{w(end)}$ After Bump	$Kr_{w(end)}$	$Kr_{o(end)}$	Average S_w Before Bump
1	Primary Drainage	2	0.20	---	0.72	---
2	Imbibition	2	0.63	0.174	---	---
3	Drainage	2	0.21	---	0.72	---
4	Imbibition	1	0.60	0.085	0.725	0.60
5	Drainage	1	0.19	---	0.721	0.22

Table 4. Displacement Experiments Conducted on Core No. 3

Experiment No.	Experiment Type	Rate cc/min	$S_{w(end)}$ After Bump	$Kr_{w(end)}$	$Kr_{o(end)}$	Average S_w Before Bump
1	Primary Drainage	2	0.14	---	0.72	---
2	Imbibition	2	0.55	0.077	---	---
3	Drainage	0.2	0.13	---	0.625	0.27

were conducted so that their data could be compared with displacement experiments.

The range of $Lv\mu$ was 2.3-9.5 Cm²/min-cp for the displacement experiments conducted on Cores No. 1 and 2, and 0.5-5 Cm²/min-cp for the experiments conducted on Core 3. This provides experimental data for a wide range of capillary effects.

The imbibition and drainage capillary pressure curves were measured for two samples from the same outcrop as Core No. 1. One of the samples was used without semi-permeable plate for the measurement of primary and secondary drainage. The maximum acceleration subjected to this core sample was corresponding to 20 psi capillary pressure. The other rock sample was used with a semi-permeable plate in the centrifuge drainage bucket and was subjected to a maximum rotational speed corresponding to 4.5 psi and then the speed was decreased in steps to trace the imbibition curve. The time given for each step was thirty hours.

History Matching and Optimization Models

An algorithm was developed to estimate relative permeabilities from laboratory displacement experiments. The approach used in this algorithm was basically the same as the one presented by Sigmund and McCaffery (1979), and Batycky et al (1981). In this work, a different method was used for the minimization of least square errors between observed and calculated values of pressure drop and recovery.

The algorithm assumes the same exponential functional form for relative permeabilities. The following functions represent relative permeabilities to oil and water:

$$K_{rnw}(Sw_D) = K_{rnw_{end}} \left(\frac{Sw_D^{E_{nw}} + A_{nw}\, Sw_D}{1 + A_{nw}} \right) \tag{1}$$

$$K_{rw}(Sw_D) = K_{rw_{end}} \left(\frac{(1-Sw_D)^{E_w} + A_w\,(1-Sw_D)}{1 + A_w} \right) \tag{2}$$

where

$$Sw_D = \frac{Sw - Swir}{1 - Sor - Swir} \tag{3}$$

A fully implicit finite difference model was developed to solve the equations describing one dimensional, horizontal two phase flow with capillary effects. These equations are:

$$\frac{K_{abs}}{\mu_w}\frac{\partial}{\partial x}\left(K_{rw}\frac{\partial P_w}{\partial x}\right) = \phi\frac{\partial S_w}{\partial t} \qquad (4)$$

$$\frac{K_{abs}}{\mu_{nw}}\frac{\partial}{\partial x}\left(K_{rnw}\frac{\partial P_{nw}}{\partial x}\right) = -\phi\frac{\partial S_w}{\partial t} \qquad (5)$$

$$P_{nw} - P_w = P_c(Sw_D) \qquad (6)$$

The initial and boundary conditions, and the method of solution are shown in detail by Sigmund and McCaffery (1979) and Settari and Aziz (1979).

In this study four parameters in relative permeability functions ($K_{rw_{end}}$, $K_{rnw_{end}}$, E_w, E_{nw}) are determined so that the following least square objective function is minimized:

$$J(K_{rw_{end}}, K_{rnw_{end}}, E_w, E_{nw}) =$$

$$\sum_{i=1}^{N}\left|\frac{\Delta p_{obs} - \Delta p_{cal}}{\Delta p_{in}}\right|^2 + \sum_{i=1}^{N}\left|\frac{R_{obs} - R_{cal}}{R_t}\right|^2 \qquad (7)$$

where Δp_{obs} and R_{obs} are measured pressure drop and recovery for ith point in a total of N data points. Also, Δp_{cal} and R_{cal} are the corresponding values of pressure drop and recovery calculated by the finite difference simulator and relative permeabilities derived from the assumed parameters (i.e, $K_{rw_{end}}$, $K_{rnw_{end}}$, E_w, E_{nw}).

If we use the following definitions,

$$Y_p = \left(\frac{\Delta P_{obs} - \Delta P_{cal}}{\Delta P_{in}}\right) \qquad (8)$$

$$Y_R = \left(\frac{R_{obs} - R_{cal}}{R_t}\right) \qquad (9)$$

$$\mathbf{X} = [K_{rw_{end}}, K_{rnw_{end}}, E_w, E_{nw}]^T \qquad (10)$$

then the objective function becomes

$$J(x) = \sum_{i=1}^{N} (Y_p^2 + Y_R^2) \tag{11}$$

The Levenberg-Marquardt minimization algorithm was used for parameter estimation. Reklaitis, Ranindran and Ragsdell (1983) show that in this method the search direction is given by:

$$S(\mathbf{X}^{(k)}) = -[\mathbf{H}^{(k)} + \lambda^{(k)}\mathbf{I}]^{-1}\nabla J(\mathbf{X}^{(k)}) \tag{12}$$

where $\mathbf{I}$ is the identify matrix; and $\mathbf{H}$ is the Hessian matrix given by:

$$H = \begin{matrix} \dfrac{\partial^2 J}{\partial x_1^2} & \dfrac{\partial^2 J}{\partial x_1 \partial x_2} & \dfrac{\partial^2 J}{\partial x_1 x_3} & \dfrac{\partial^2 J}{\partial x_1 x_4} \\ \cdot & \cdot & \cdot & \cdot \\ \cdot & \cdot & \cdot & \cdot \\ \cdot & \cdot & \cdot & \cdot \\ \dfrac{\partial^2 J}{\partial x_1 \partial x_4} & \dfrac{\partial^2 J}{\partial x_2 \partial x_4} & \dfrac{\partial^2 J}{\partial x_3 \partial x_4} & \dfrac{\partial^2 J}{\partial x_4^2} \end{matrix} \tag{13}$$

$$\nabla J = \begin{matrix} 2\sum_{i=1}^{N} \left| Y_p \dfrac{\partial Y_p}{\partial x_1} + Y_R \dfrac{\partial Y_R}{\partial x_1} \right| \\ \vdots \\ 2\sum_{i=1}^{N} \left| Y_p \dfrac{\partial Y_p}{\partial x_4} + Y_R \dfrac{\partial Y_R}{\partial x_4} \right| \end{matrix} \tag{14}$$

and λ is the Marquardt search direction and step length parameter. When λ approaches $+\infty$, the Marquardt's algorithm reduces to the method of steepest descent. When λ is equal to zero, the algorithm reduces to the Gauss-Newton.

RESULTS AND DISCUSSION

The measured drainage and imbibition capillary pressure curves are shown in Figure 3. Although these curves are from two different samples, we assumed that they are applicable for simulation purposes.

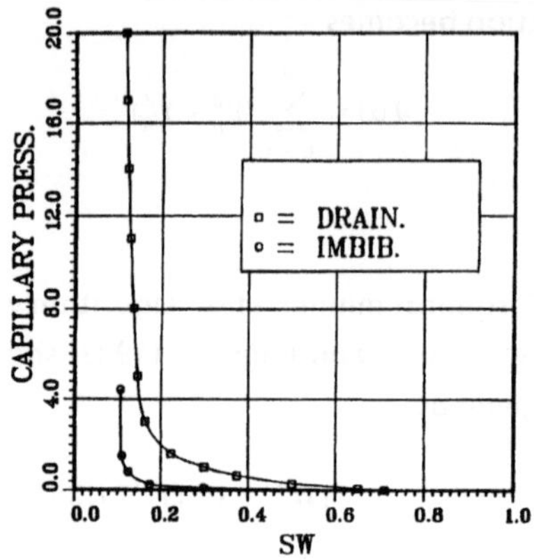

Fig. 3. **Capillary pressure curves for Berea sandstone core.**

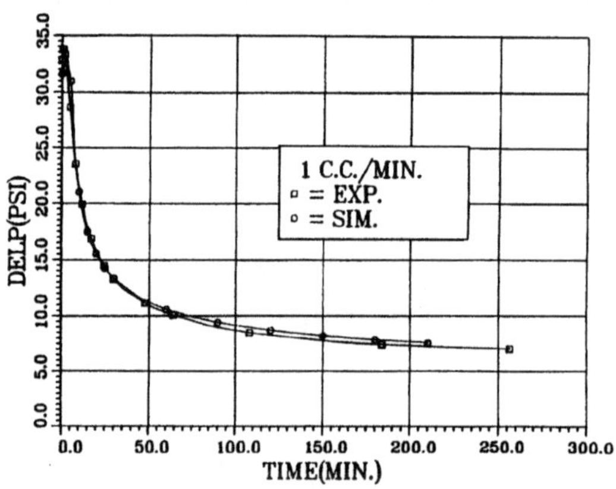

Fig. 4. **Comparison of experimental and history-matched pressure drop data for 1 cc/min run.**

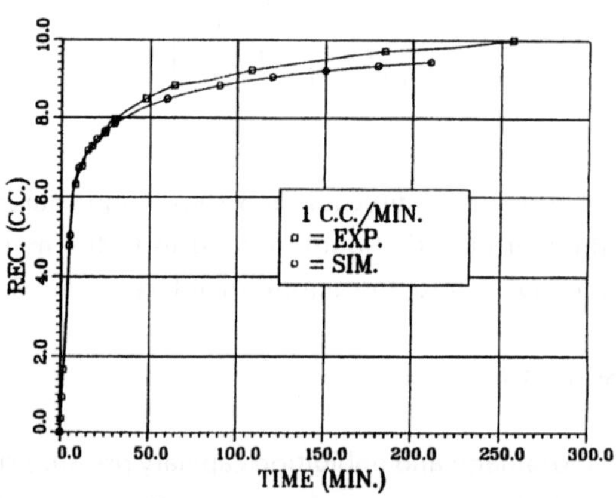

Fig. 5. **Comparison of experimental and history-matched recovery data for 1 cc /min run.**

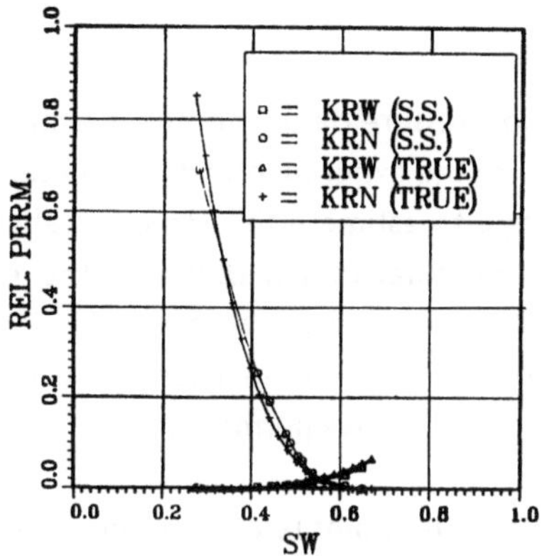

Fig. 6. Steady-state and "true" drainage relative permeability curves for Core No. 1.

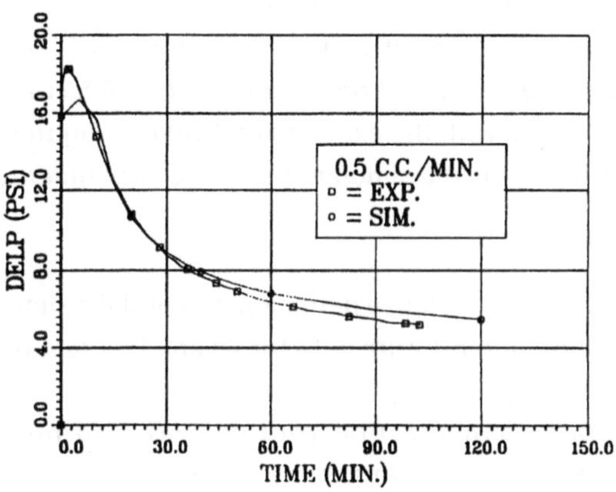

Fig. 7. Experimental and simulated pressure drop data for 0.5 cc/min.

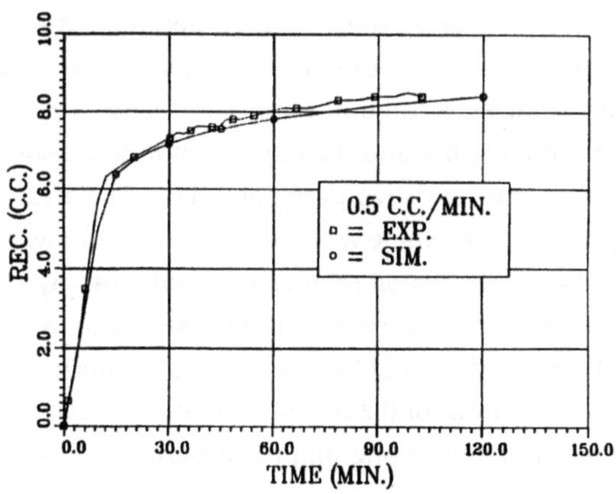

Fig. 8. Experimental and simulated recovery data for 0.5 cc/min.

Drainage

The production and pressure data from the 1 cc/min drainage run in Core No. 1 were used along with the experimental drainage capillary pressure curve in the history match package to estimate the "true" relative permeability curves. Figures 4 and 5 show the final match between the experimental and the calculated pressure and recovery values respectively. The final error for this match was 0.108. Figure 6 shows a comparison between the measured steady state and estimated or "true" relative permeabilities.

Although steady state end point relative permeabilities are lower than the calculated "true" relative permeability end-points at the mid-saturation range relative permeabilities to oil by steady state are higher than the "true" oil relative permeability. This may be due to different fluid distributions occurring in these two types of displacement (Handy and Datta, 1966). To further evaluate the reliability of the "true" relative permeabilities, they were used in the simulator to calculate recovery and pressure data from 0.5 cc/min displacement in Core No. 1, to compare with the experimental values. Figures 7 and 8 show a satisfactory match between the simulated and the experimental pressure and recovery data.

At this stage the "true" relative permeabilities and the simulator were used to generate production and pressure data for several low rate displacements. A two phase non-capillary Buckley Leverett algorithm was developed. The Marquardt-Levenberg minimization routine was incorporated with this algorithm to estimate relative permeability parameters for the minimum least square error between the relative permeabilities from the capillary two phase flow simulator and the Buckley Leverett model. Since Buckley Leverett solution does not consider the capillary effects, these relative permeabilities are wrong and we call them "false". For all these relative permeabilities we assume that the end-point saturations are constant and equal to "true" end-point saturations. Figure 9 shows a comparison of the "false" relative permeability curves calculated from 0.05, 0.1 and 0.2 cc/min simulated displacement runs in Core No. 1 with the "true" relative permeabilities. The decrease in relative permeabilities from the "true" value with decrease in rate show a definite trend. This indicates that the calculated oil relative permeabilities by JBN or Jones and Roszelle for low flow rate displacement experiments are in large errors. Figure 10 shows that the functional form of the "false" relative permeabilities found by parameter estimation method for 0.2 cc/min simulated production data, is a very accurate representation of what one might calculate from Jones and Roszelle method.

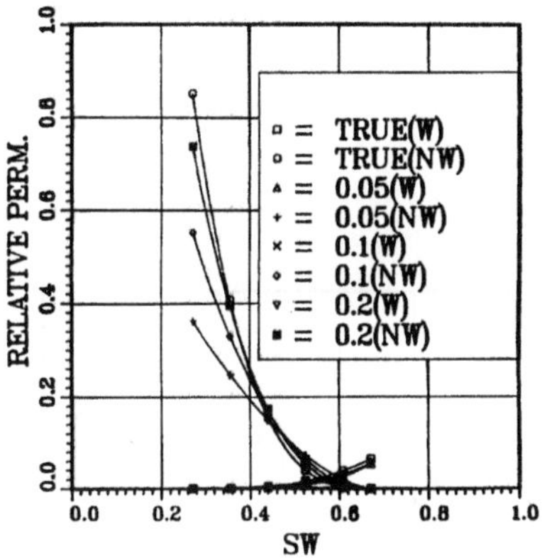

Fig. 9. "True" and "false" relative permeability curves for different rates.

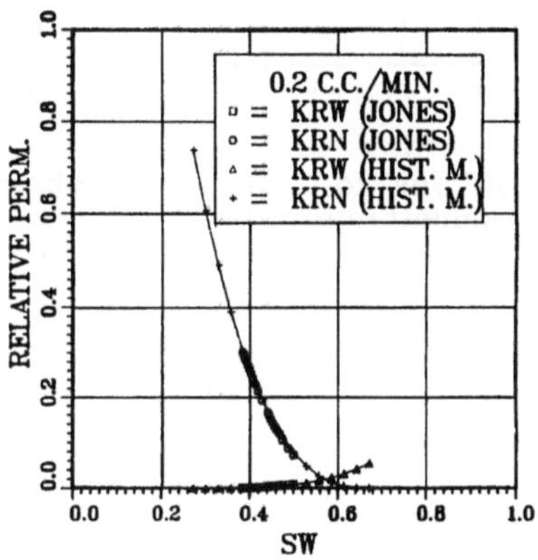

Fig. 10. Comparison of relative permeability curves generated by history matching and Jones and Roszelle method.

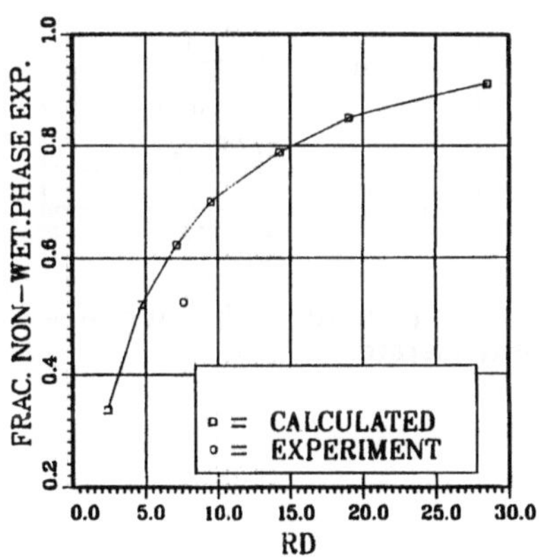

Fig. 11. Ratio of "true" to "false" non-wetting phase exponents vs. dimensionless rate.

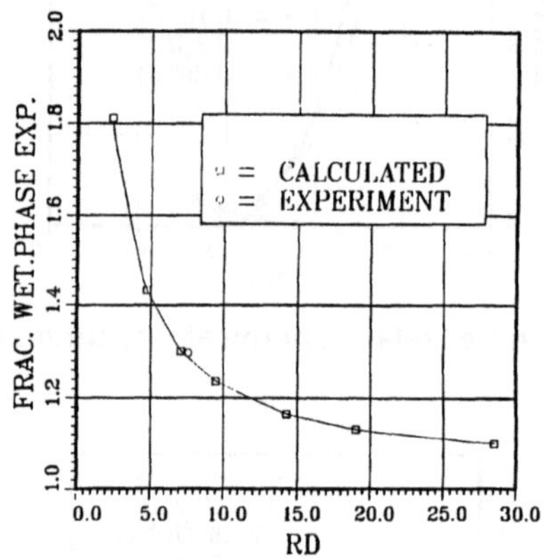

Fig. 12. Ratio of "true" to "false" wetting phase exponents vs. dimensionless rate.

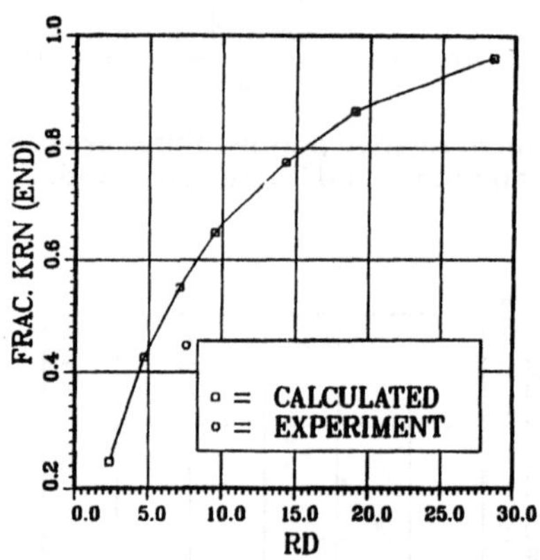

Fig. 13. Ratio of "true" to "false" endpoint non-wetting phase relative permeability vs. rate.

Since the relative permeabilities are defined by the parameters (E_{nw}, E_{w}, $Kr_{nw_{end}}$, $Kr_{w_{end}}$), the change in relative permeabilities with rate can now be quantified with the change in parameters. Figures 11, 12, 13, and 14 show the above parameters as a function of rate. On the y-axis of each one of these figures, the value of each parameter is divided by the value of the same parameter from "true" relative permeability function. The x-axis in all these figures is dimensionless rate which is defined as:

$$R_D = \frac{L\mu q}{AK\left(\frac{\sigma}{\sqrt{K/\phi}}\right)} \tag{15}$$

Regression equations both in exponential and polynomial formats were fit to the curves plotted on Figures 11 through 14 to find the functional form of the change of each parameter with dimensionless rate. The following are the functions:

$$\frac{Kr_{nw_{end}}}{(Kr_{nw_{end}})_{true}} = 0.188 - 1.4 \times 10^{-5} exp\,(R_D/2) + 0.045\,R_D \tag{16}$$

$$\frac{E_{nw}}{(E_{nw})_{true}} = 0.308 - 1.4 \times 10^{-5} exp\,(R_D/2) + 0.038\,R_D \tag{17}$$

and

$$\frac{E_{w}}{(E_{w})_{true}} = 1.28 + 1.8\,exp\,(-R_D/2) - 0.0071\,R_D \tag{18}$$

for

$$\frac{Kr_{w_{end}}}{(Kr_{w_{end}})_{true}}$$

the exponential format was not applicable.

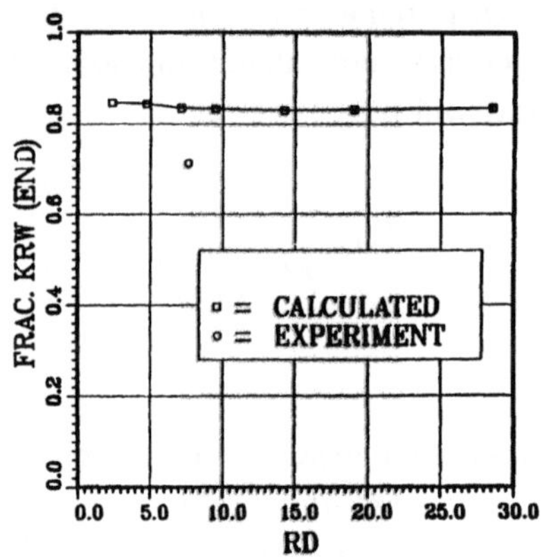

Fig. 14. Ratio of "true" to "false" endpoint wetting phase relative permeabilities vs. rate.

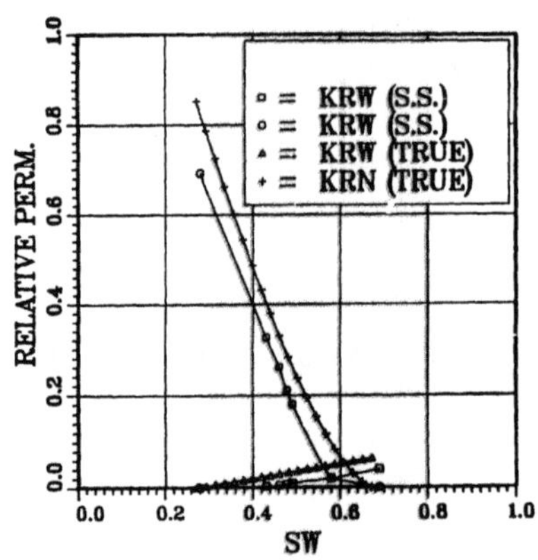

Fig. 15. Comparison of steady-state and "true" imbibition relative permeabilities.

Polynomial format:

$$\frac{Kr_{nw_{end}}}{(Kr_{nw_{end}})_{true}} = 0.14 + 0.062\,R_D - 1.2 \times 10^{-3} R_D^2 \tag{19}$$

$$\frac{Kr_{w_{end}}}{(Kr_{w_{end}})_{true}} = 0.85 - 0.0025\,R_D - 7 \times 10^{-5} R_D^2 \tag{20}$$

$$\frac{E_{nw}}{(E_{nw})_{true}} = 0.705 - 0.95\,R_D^{-1} + 9 \times 10^{-4} R_D \tag{21}$$

$$\frac{E_{w}}{(E_{w})_{true}} = 1.04 + 1.82\,R_D^{-1} - 4.6 \times 10^{-4} R_D \tag{22}$$

To investigate the applicability of these functions to other cores, the displacement experiments in other core samples were analyzed.

Cores No. 2 and 3 were from the same outcrop with the same approximate permeability. We assumed that the capillary effects in the 2 cc/min run from Core No. 2 which was the longer core in practically negligible. The production and pressure data from this experiment was used in the Buckley Leverett model to estimate the relative permeability parameters. Also the 0.2 cc/min run production data from Core No. 3 were analyzed.

The ratio of parameters from these two experiments are shown as the experimental point on Figures 11 to 14. This shows that although there is a functional relationship between the parameters and rate, the application of these functions to correct the relative permeabilities for another core is questionable. Further investigation on experimental verification of these relationships is essential.

Imbibition

The "true" imbibition relative permeability function was found by using the imbibition capillary pressure, and the recovery and pressure data from 1 cc/min displacement run of Core No. 1. Figure 15 shows a comparison of "true" and steady state relative permeability curves. The "true" relative permeabilities are higher than steady state relative permeabilities at the end points and the mid saturation range. A lower curvature is observed for "true" relative permeability curve. We tried to follow the same procedure to find any functional relationship between the parameters of imbibition relative permeability and rate, but the final match did not result to any meaningful shape of relative permeabilities at the lower rates.

CONCLUSIONS

A computer model was developed to estimate the oil and water relative permeability exponential functions from displacement experiments. The model consist of an optimization algorithm, a one-dimensional two phase flow simulator with capillary effects, and a simple Buckley Leverett non-capillary flow simulator. Using experimental data and the model, the effect of rate on the parameters of relative permeability functions were studied. The following conclusions can be delivered from the results of this study.

1. In the case of drainage, the end-point relative permeability and saturation exponent for the non-wetting phase can be related to the dimensionless rate. These parameters increase as the dimensionless rate increases.

2. The saturation exponent in the wetting phase relative permeability functional form decreased with an increase in rate.

3. No significant change was observed between the end-point relative permeability to the wetting phase rate.

4. The results of the end-point relative permeability to the wetting phase did not show any significant change with rate.

5. In the case of imbibition no meaningful rate dependent trend was observed for the end-point relative permeability and saturation exponent.

ACKNOWLEDGEMENT

This work was part of the senior author's thesis project.

Financial support provided by the Petroleum Development Laboratory and the Petroleum Engineering Department at the University of Alaska is gratefully acknowledged.

NOMENCLATURE

A = area
Anw = constant in non-wetting phase relative permeability expression
A_w = constant in wetting phase relative permeability expression
E_{nw} = non-wetting phase parameter in relative permeability expression
E_w = wetting phase parameter in relative permeability expression
i = injection rate
J = least squares error function to be minimized
K = absolute permeability
Kr = relative permeability
Kr_{nw} = relative permeability of non-wetting phase
Kr_w = relative permeability of wetting phase
$Krnw_{end}$ = end-point relative permeability of non-wetting phase
Krw_{end} = end-point relative permeability of wetting phase
L = length
N = number of data point observations
Δp = pressure drop across the core
Δp_{calc} = calculated Pressure drop across the core
Δp_{obs} = observed pressure drop across the core
Δp_{in} = initial pressure drop across the core
P_{nw} = pressure in non-wetting phase
P_w = pressure in wetting phase
R_{calc} = recovery calculated from simulator
R_{obs} = recovery observed experimentally
R_t = total recovery
S_w = wetting phase saturation
$(S_w)_{init}$ = initial wetting phase saturation
$(S_w)_{min}$ = minimum wetting phase saturation
$(S_w)_{max}$ = maximum wetting phase saturation
S_{wir} = irreducible wetting phase saturation
t = time
x = space coordinate
X = vector of parameters in Marquardt algorithm (see Equation 10)
ϕ = porosity
λ = Levenberg-Marquardt parameter
μ = viscosity
σ = interfacial tension

SUPERSCRIPTS

k = iteration in Marquardt algorithm

end = end-point designation for relative permeability

$\bar{}$ = average property

SUBSCRIPTS

cal = calculated

i = index of observation

ir = irreducible

in = initial

min = minimum

max = maximum

nw = non-wetting phase

obs = observed

w = wetting phase

REFERENCES

1. Archer, J.S., and Wang, S.W.: *"Use of a Reservoir Simulator to Interpret Laboratory Water Flood Data"*, Soc. Pet. Eng. J. (Dec. 1973) 363-367.

2. Aziz, K., and Settari, A.: Petroleum Reservoir Simulation, Elsevier Applied Science Publishers, London, 1979.

3. Batycky, J.P., McCaffery, F.G., Hodgins, and P.K., Fisher, D.B.: *"Interpreting Relative Permeability and Wettability From Unsteady-State Displacement Measurements"*, SPE J. (June 1981) 296-308.

4. Chouke, R.L., Van Meurs, P., and Vander Poel, C.: *"The Instability of Slow, Immiscible, Viscous Liquid - Liquid Displacements in Permeable Media"*, Trans. AIME, 216, 188-194 (1959).

5. Civan, F., and Donaldson, E.C.: *"Relative Permeability From Unsteady State Displacements: An Analytical Interpretation"*, SPE 16200 presented at SPE Production Operations Symposium, Oklahoma City, OK (March 8-10, 1987).

6. Engleberts, W.F. and Klinkenberg, L.J.: *"Laboratory Experiments on the Displacement of Oil by Water From Packs of Granular Materials"*, proc. 3rd World Pet. Congr., The Hague, part II, 544 (1951).

7. Gabriel, G.A. and Inamdar, G.R.: *"An Experimental Investigation of Fines Migration in Porous Media"*, SPE 12168, presented at the 58th Annual Technical Conference of SPE, San Francisco, CA, (Oct. 5-8, 1983).

8. Handy, L.L. and Datta, P.: *"Fluid Distribution During Immiscible Displacements in Porous Media"*, Trans. AIME (1966) 237, 261.

9. Heaviside, J. and Black, C.J.J.: *"Fundamentals of Relative Permeability: Experimental and Theoretical Considerations"*, SPE 12173 presented at the 58th Annual Technical Conference of SPE-AIME, San Francisco (October 5-8, 1983).

10. Johnson, E.F., Bossler, D.P. and Naumann, V.O.: *"Calculation of Relative Permeability from Displacement Experiments"*, Trans. AIME (1959) 216, 370-372.

11. Jones, S.C. and Roszelle, W.O.: *"Graphical Techniques for Determining Relative Permeability From Displacement Experiments"*, J. P. T. (May 1978) 807-817.

12. Kerig, P.D., and Watson, A.T.: *"A New Algorithm for Estimating Relative Permeability From Displacement Experiments"*, SPE 14476 unsolicited, (April 29, 1985).

13. Kyte, J.R., and Rapopart, L.A.: *"Linear Water Flood Behavior and End Effects in Water-Wet Porous Media"*, Trans. AIME (1958) 213, 423-426.

14. Odeh, A.S, and Dotson, B.J.: *"A Method for Reducing the Rate Effect on Oil and Water Relative Permeabilities Calculated From Dynamic Displacement Data"*, SPE 14417 presented at the 60th Annual Technical Conference of the SPE, Las Vegas, NV (Sept. 22-25, 1985).

15. Peters, E.J., and Flock, D.L.: *"The Onset of Instability During Two-Phase Immiscible Displacements in Porous Media"*, SPE J. (April 1981) 249-258.

16. Rapopart, L.A., and Leas, W.A.: *"Properties of Linear Water Floods"*, Trans. AIME, 198, 139 (1959).

17. Reklaitis, G.V., Ravindran, A., and Ragsdell, K.M.: Engineering Optimization Methods and Applications, John Wiley & Sons, New York City (1983).

18. Saraf, D.N., and McCaffery, F.G.: *"Two and Three-Phase Relative Permeabilities: A Review"*, Petroleum Recovery Institute Report 1981-8, (Sept. 1981), 28.

19. Sigmund, P.M. and McCaffery, F.G.: *"An Improved Unsteady - State Procedure for Determining the Relative Permeability Characteristics of Heterogeneous Porous Media"*, Soc. Pet. Eng. J. (Feb. 1979) 15-28.

20. Tao, T.M., and Watson, A.T.: "Accuracy of JBN Estimates of Relative Permeability: Part 1 - Error Analysis", SPE J. (April 1984) 209-214.

21. Tao, T.M., and Watson, A.T.: *"Accuracy of JBN Estimates of Relative Permeability: Part 2 - Algorithms"*, SPE J. (April 1984) 215-222.

SHALLOW SANDS OF NORTH SLOPE, ALASKA AND THEIR HYDROCARBON POTENTIALS

G. D. Sharma, D. O. Ogbe, V. A. Kamath and M. Zhang

Petroleum Development Laboratory
University of Alaska Fairbanks
Fairbanks, AK

ABSTRACT

Extensive shallow sand deposits on North Slope Alaska contain about 35 billion barrels of oil in place. These Cretaceous and Tertiary sands are divided into two distinct units; the upper unit is termed as Ugnu sands and the lower unit is known as West Sak sands.

The Ugnu sands which occur at 2,000 ft below surface in the southwest dip towards northeast to a depth of 4,000 ft. These sands are divided into upper and lower units. The upper unit is discontinuous while the lower Ugnu sands are on an average 275 ft thick with good reservoir properties. The units contain approximately 15 billion barrels of 8-12°API crude characterized as tar sand bitumen.

The underlying West Sak sands, at depth ranging from 2,500 to 4,500 ft, consist of upper and lower units. These units contain approximately 20 billion barrels of crude ranging from 10 to 22°API.

These reservoirs are described using well log, core description and analyses of fluids therein. The well log interpretation was done using LOGCALC. Various reservoir parameters were evaluated to (i) characterize each pay zone, (ii) correlate lithofacies distributions (iii) estimate hydrocarbon in place, and (iv) environment of deposition.

The viscosities of crude oil found in these sands are typically high because of low temperature and recovery from these sands will require enhanced oil recovery (EOR) technology. Based on the EOR screening, miscible and thermal processes are considered to be suitable for recovery of crude from West Sak sands, while only thermal processes are considered to be applicable for recovery of Ugnu tar sands. The unconsolidated and friable sands also pose problems in well completion and production.

INTRODUCTION

About 100 billion barrels of heavy oil have been discovered in the United States, and only about 10% of that has been produced so far. It is estimated that

Particle Technology and Surface Phenomena in Minerals and Petroleum
Edited by M.K. Sharma and G.D. Sharma, Plenum Press, New York, 1991

half of the total heavy oil deposits (51 billion bbl) in the U.S. cannot be recovered with the available technology. Much of this unrecoverable heavy oil either is in fields which lie at depths greater than 3,000 ft or is too viscous to recover using simple thermal recovery (steam injection). However, some of it may be recovered through injections of steam with additives.

California, with the largest accumulations of heavy oil estimated at 42 billion bbl, has recovered approximately 10 billion bbl of heavy oil. Alaska which contains the second largest heavy oil deposits has about 20 billion bbl. Extensive deposits of tar sands containing nearly 15 billion bbl of hydrocarbons also lie in Alaska. The heavy oil occurs in shallow sands of super giant West Sak Field, and tar occurs in the overlying sands of Ugnu Field. It is estimated that over 6 billion bbl of oil can be recovered from the West Sak Field alone by using known and advanced recovery technologies such as steam injection through horizontal wells, enriched gas miscible, steam additives, and in-situ combustion. The recovery of hydrocarbons from Ugnu sands probably will require specialized recovery techniques.

Prior to the selection of effective EOR processes for West Sak and Ugnu reservoirs, it is essential to study the reservoir parameters and their spatial distributions in detail. For commercial production, a thorough understanding of reservoir framework, architect and characterization is of the utmost importance. The purpose of this paper is to describe these parameters for West Sak and Ugnu reservoirs and evaluate their oil recovery potentials.

GEOLOGY OF NORTH SLOPE

The major oil fields of Alaska lie in the northern part of the state. The narrow strip called the North Slope lies between the Brooks Range in the south and the Beaufort Sea coast in the north (Figure 1). Giant fields and a few smaller field discovered and currently producing on the North Slope include Prudhoe Bay, Kuparuk River, West Sak, Lisburne, Milne Point, and Endicott. To the west of these producing fields lies the National Petroleum Reserves - Alaska (NPR) while the Arctic National Wildlife Refuge (ANWR) is located to the east. North Slope production is divided primarily into two major units; the Prudhoe Bay Unit and the Kuparuk River Unit (Figure 2). The Prudhoe Bay Unit production is from deeper and older formations while Kuparuk River Unit production is from shallower reservoirs with low API and low GOR.

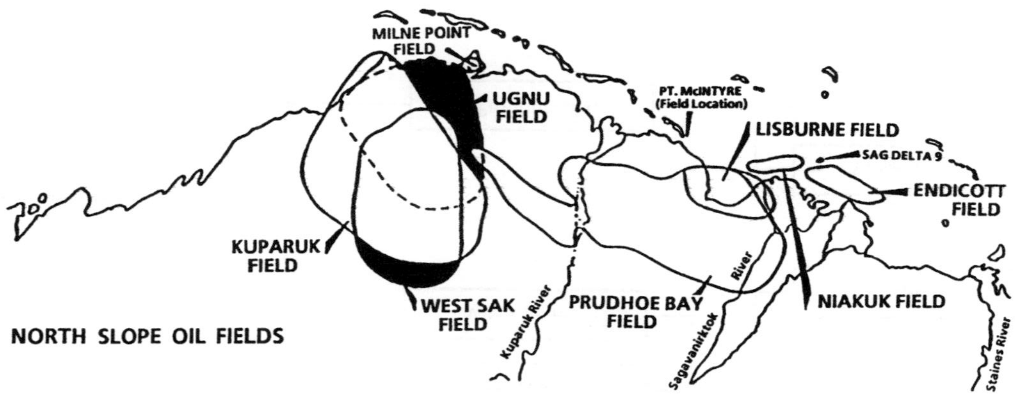

Figure 1. Map Illustrating Locations of the Major Reservoirs On the North Slope (Source: BP America, 1989)

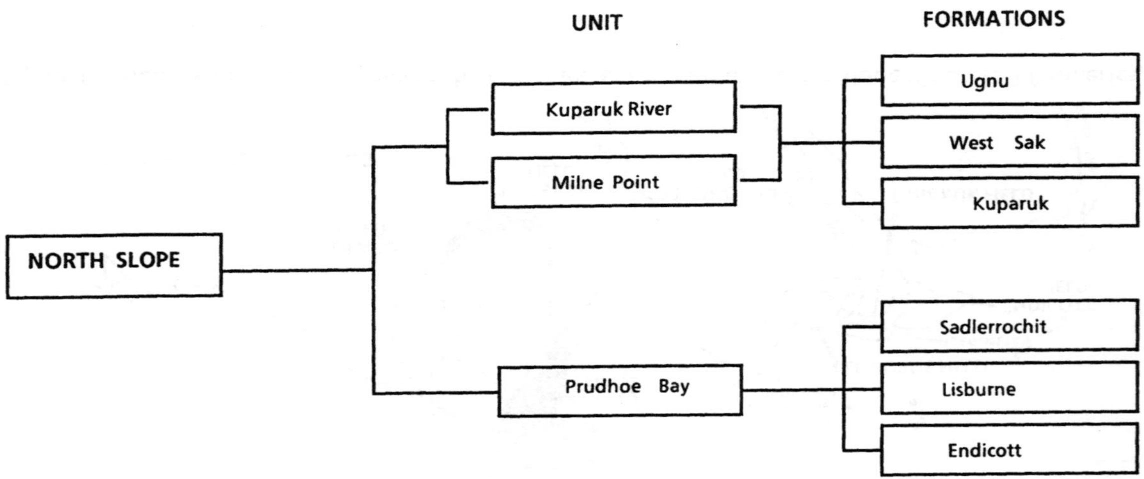

Figure 2. Producing Formations of Various Units in North Slope

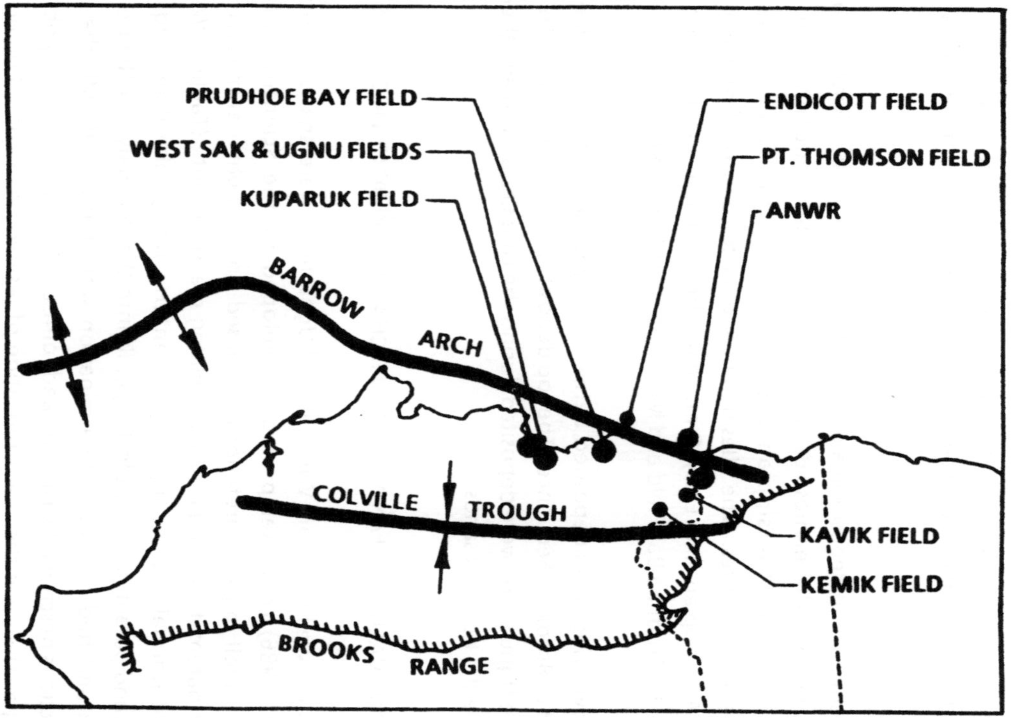

Figure 3. Alaska North Slope Basin Showing Barrow Arch, Colville Trough and Brooks Range

The accumulation of vast quantities of oil, gas and tar along the North Slope is primarily due to the large amounts of sediments deposited in the east-west oriented Colville Trough, which was formed during Early Paleozoic Era. The thickness of sediments in this large synclinal trough varies from 25,000 ft to over 30,000 ft in the deepest part. The sediments deposits include thick sequences of shales which served as source rock as well as thick sandstone units which serve as reservoirs for large pools of hydrocarbons. The dominant structures for controlling migration and trapping are a broad submerged east-west oriented Barrow Arch along the Beafort Sea coast, erosional unconformities and faulting. (Figure 3)

The sequence of deposition along the North Slope can be divided into three major sequences. The earliest deposits of Ellesmerian Sequence overlie the nonproductive Pre-Mississippian basement rocks consisting of quarzitic greywackes and argillites. The Ellesmerian Sequence consists of rock deposited between Mississippian and Early Cretaceous time. These include productive Permo-Triassic Prudhoe Bay Field and Mississippian Lisburn Field.

The overlying rocks deposited between Early Cretaceous and Tertiary time are termed as Brookian Sequence. The top deposits are of Cenozoic Era. The West Sak and the Ugnu sands were deposited during the Late Cretaceous and therefore occur above the Kuparuk sands.

FORMATION EVALUATION AND RESERVOIR CHARACTERIZATION METHODS

Available well logs from West Sak and Ugnu fields were analyzed using the *LOGCALC*[1] (1986) computer program developed by the Scientific-Software Intercomp. Well log interpretations are based on petrophysical models which relate rock parameters to a particular well log measurement. (Table I) Well log traces were normalized by applying the interactive graphical techniques of histogram and parameter cross-plotting built into the log analysis software. Data smoothing and bed boundary determination were performed with the aid of Walsh functions described by Lanning and Johnson[2] (1983). The primary objective of well log analysis was the characterization of reservoir rocks based on porosity, water saturation and other parameters. The log-derived data was then analyzed to evaluate the potential for hydrocarbon in these reservoirs. The reservoir quality of these shallow sands was evaluated based on certain minimum reservoir parameters. The net pay zones were determined by the following cut-off values:

Table I. Flow Chart for Well Log Analysis and Reservoir Characterization

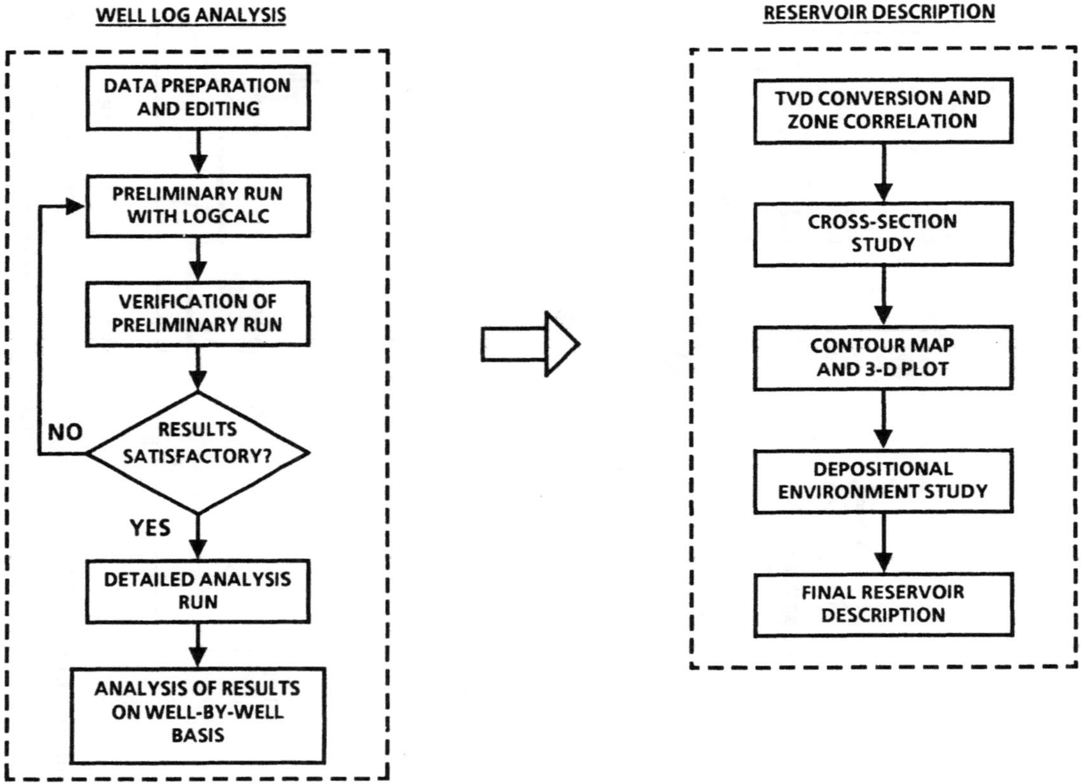

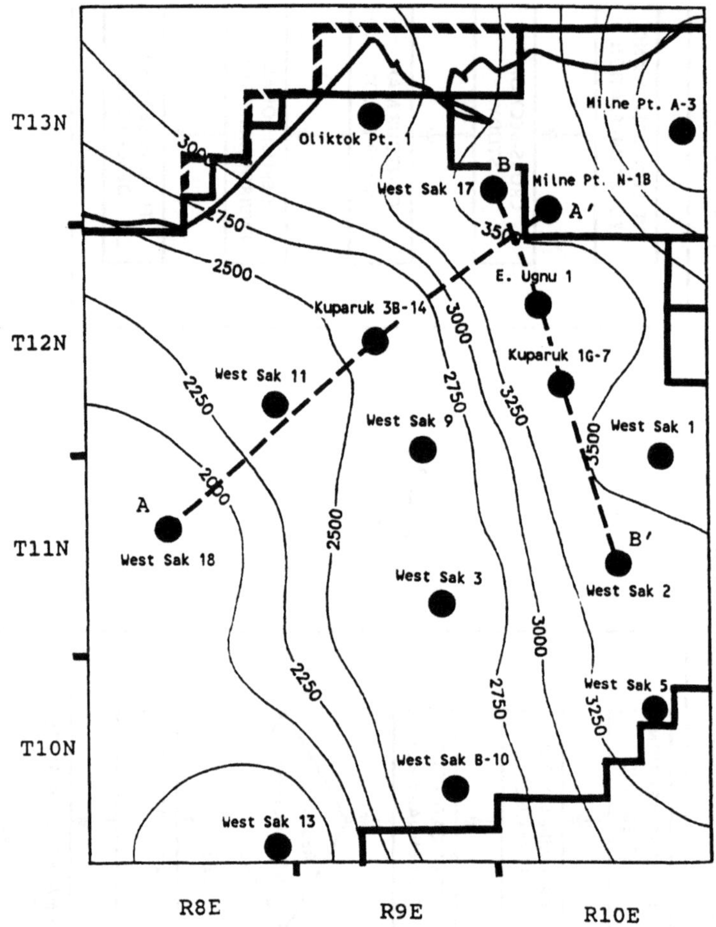

Figure 4. Contour Map Showing Depths to the Top of West Sak Upper Sand 1 (AA': SW-NE Cross-Section, BB': NW-SE Cross-Section)

porosity	< 6%
water saturation	> 50%
shale volume	> 40%

WEST SAK FORMATION

The West Sak Reservoir, discovered in 1971, consists of fine to fine-grained sands and silty sands interbedded with silts and muds deposited during Late Cretaceous Period. The sand consists primarily of quartz, lithic rock and feldspar with minor components of clays with traces of mica and glauconite. The sands are semi-consolidated to unconsolidated. The thickness of West Sak sands varies from 450 ft in the southwest to 200 ft in the northeast. In the southwest they occur at 2,300 ft below the surface and dip gently (1° - 2°, 130 ft/mile) to a depth of 4,100 ft in the northeast. The West Sak sands form a monocline with strike to the northeast (Figures 4, 5 and 6). A few major and a number of minor north-south trending faults with mostly downthrown to the east cut the entire sands and further descend into underlying Kuparuk River Formation of Ellesmerian Sequence. The throw on these faults may range from 50 ft to 150 ft. These faults and the stratigraphy (general dip) in various hydrocarbon bearing sands form traps for hydrocarbons accumulation.

West Sak Field contains over 20 billion barrels of oil in place and this is the largest heavy oil reservoir in North America. The giant reservoir covering over 300 square miles (775 km^2) lies mostly over the Kuparuk River and Milne Point fields in North Slope, Alaska. The depth of reservoir in West Sak Field varies from 2,500 ft to 4,500 ft. Because of the overlying permafrost, the temperature in the reservoir rock is quite low and varies from 45°F in shallow regions to 100°F in the deeper parts. The gravity of the oil in the reservoir is also controlled by the depth and varies from 10.5 - 22.5°API. The viscosity of the oil, to some extent, is also controlled by the degree of degradation which increases with decreasing depth of the reservoir. The West Sak oil also has a low GOR. At shallower depths, the oil in West Sak Field is quite viscous due to (1) the low temperature caused by the overlying 2,000 ft thick permafrost, (2) escape of lighter components caused by lesser overburden, and (3) increased degredation of oil by permeating surface waters.

The deeper reservoir of West Sak Field in the vicinity of Milne Point Field and Schrader Bluffs is currently under production. The gravity of oil produced varies from 14° to 20°API. About 20,000 bbl/day are produced from Milne Point Field, while at Schrader Bluffs, which is currently being developed, the daily production is anticipated at 16,500 bbl/day. The production from West Sak Reservoir from

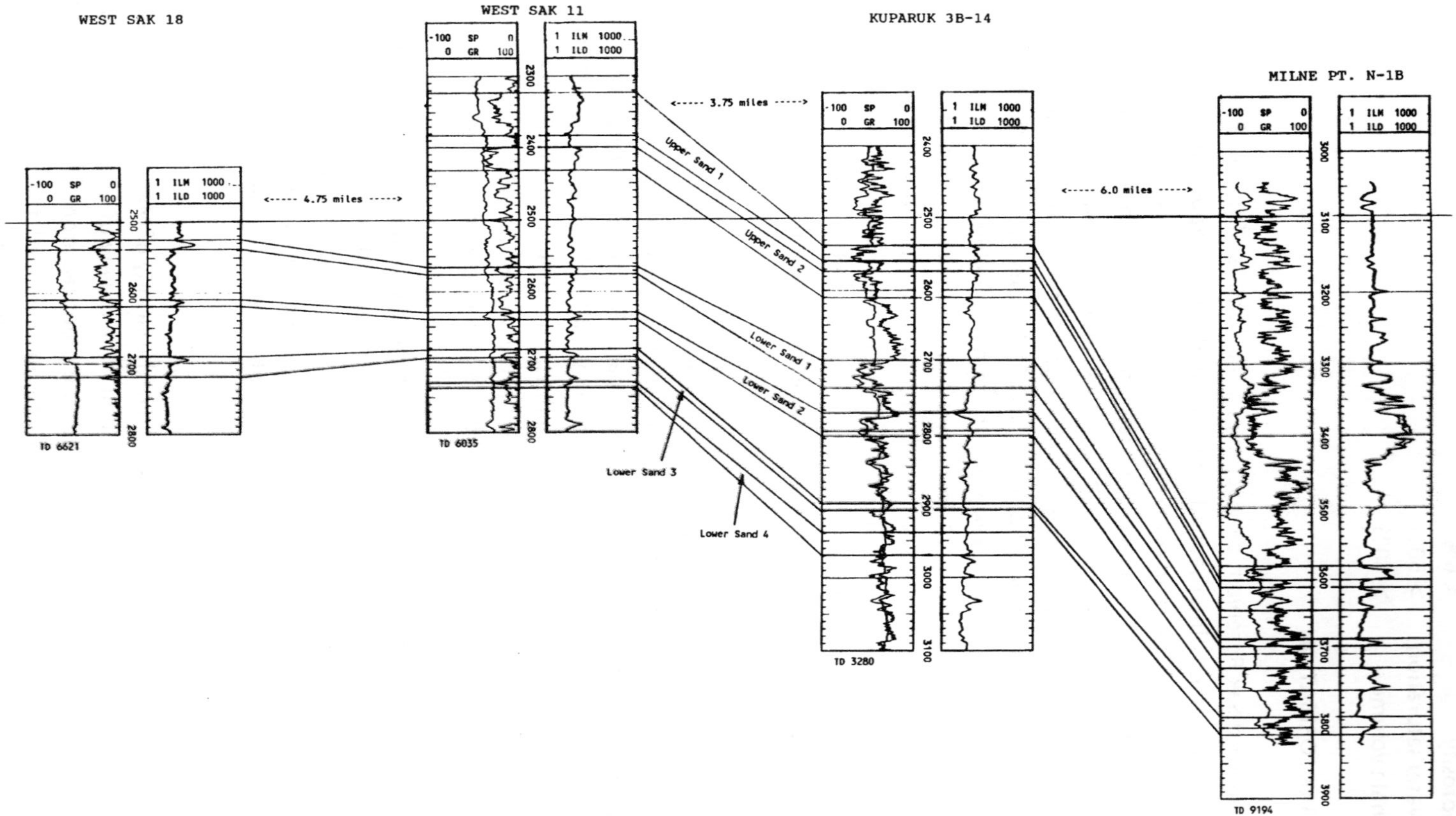

Figure 5. West Sak Sands Depicted on TheSW-NEStratigraphic Cross-Section AA'

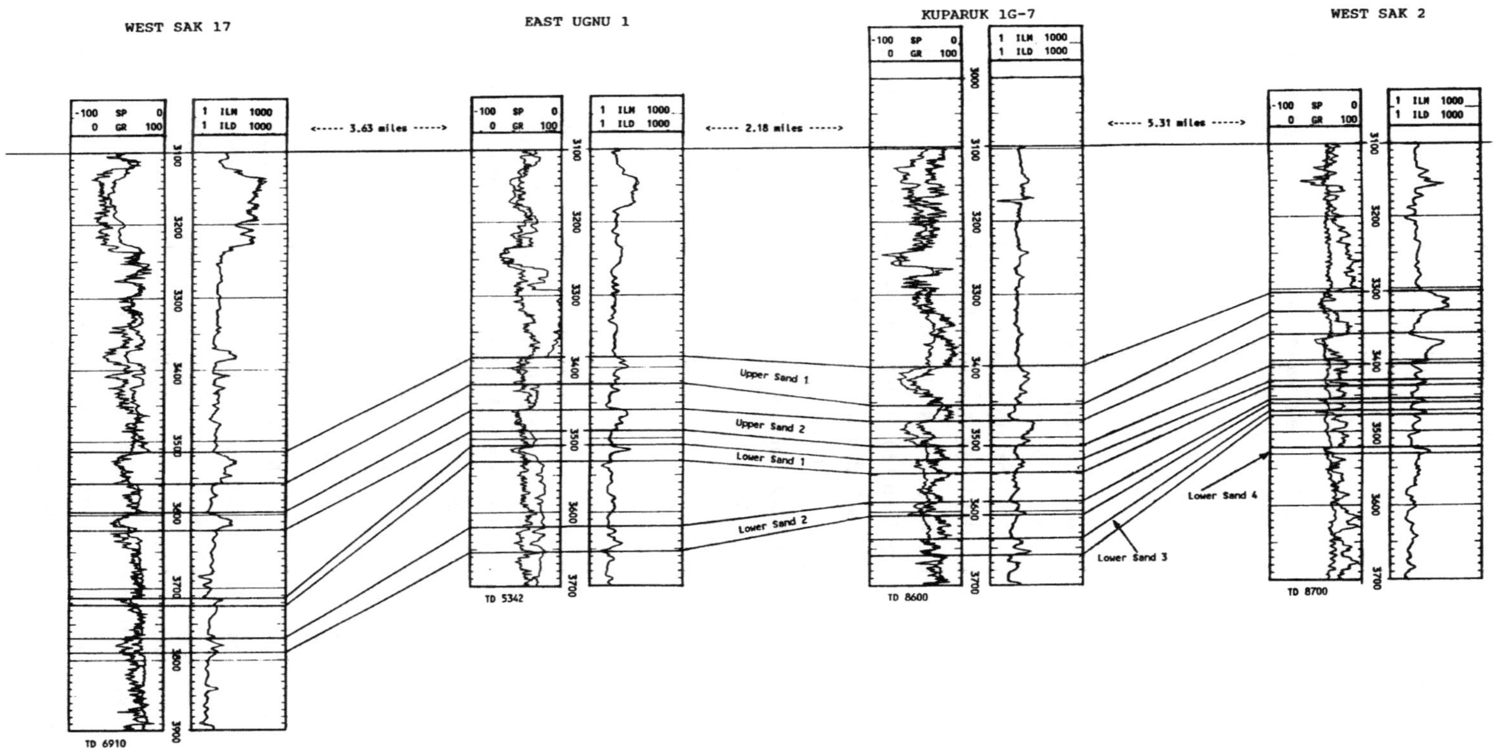

Figure 6. West Sak Sands Depicted on the NW-SE Stratigraphic Cross-Section BB '

medium and shallow depths, however, will require the use of enhanced oil recovery (EOR) processes.

Stratigraphy

The well logs from West Sak Field were analyzed to study the stratigraphy and vertical sequential deposition. Furthermore, distinct lithofacies were characterized for their reservoir potentials. The spatial distributions of each of the potential reservoir were obtained by subsurface correlation and are shown in NW-SE and SW-NE stratigraphic cross-sections (Figures 5 and 6).

West Sak deposits consist of alternate sequences of shales, shaly sands and sands. Various sand layers, however, are widely distributed and are easily correlated between the well throughout the field. These deposits can be subdivided into two units: the Upper Sands and the Lower Sands. The Lower and Upper Sands are separated by a thick shale which reaches a thickness of over 130 ft in the southwest (Figures 5 and 6). The Lower Sands can be subdivided further into four distinct lithofacies which extend laterally throughout the field. These are termed in ascending order as Lower Sand 4, Lower Sand 3, Lower Sand 2 and Lower Sand 1. Similarly the Upper Sands consist of two distinct facies. It should be noted that the Upper Sands are absent in the southwest due to truncation as a result of erosion (Figure 5).

Depositional Environment

The SP (funnel and egg-shaped) and resistivity patterns of West Sak deposits suggest that these sediments were deposited during a major regressive sea with minor eustatic fluctuations. Generally, the Lower Sands display a lesser degree of sorting and a larger percentage of fine-grained sand which is characteristic of inner shelf deposits. The Upper Sands on the other hand are coarser in grain size and contain lesser amounts of fines, generally deposited in shallow waters of a high energy environment.

It should be noted that the shale separating Lower and Upper sands thickens towards southwest, thus indicating deepening of water and shifting of the depocenter in that direction. However, the shale base in Upper Sands changes abruptly, thus indicating an hiatus in deposition and erosion. A slow northerly retreat of sea with minor sea level fluctuation should account for generally upward coarsening of sediments and eventually non-deposition and erosion of Upper Sands in the southwest region of West Sak Field. As a consequence, the

Table II. Petrophysical Properties of West Sak Lower Sands (Sand 1)

Well Name	Interval (ft)	Net Pay (ft)	Average ϕ (%)	Average Sw (%)
West Sak No. 1	3860-3874	11	35.6	38.7
West Sak No. 2	3428-3434	5	28.3	16.5
West Sak No. 3	2753-2772	-	-	-
West Sak No. 5	No log data	available		
West Sak No. 9	2852-2860	6	33.5	40.5
West Sak No. B-10	2934-2945	3	33.7	45.3
West Sak No. 11	2566-2575	-	23.9	59.3
West Sak No. 17	3714-3722	-	16.9	79.8
West Sak No. 18	2526-2536	4	36.4	35.4
East Ugnu No. 1	3511-3530	5	33.6	42.2
Kuparuk No. 1G-7	3533-3540	2	21.1	46.2
Kuparuk No. 3B-14	2698-2740	37	39.8	40.5
Milne Pt. No. N-1B	3677-3684	4	16.9	37.1

Table III. Log-Derived Data for Selected Wells in the West Sak Sands

Well Name	Interval* (ft)	Weighted Mean Effective Porosity (%)	Weighted Mean Water Saturation (%)	Net Pay Thickness (ft)	Net to Gross Ratio
West Sak No. 1	3746-4044	34.6	32.5	116	0.39
West Sak No. 2	3304-3530	28.7	19.0	84	0.37
West Sak No. 3	2628-2882	29.95	33.6	41	0.16
West Sak No. 5	3332-3428	26.8	18.0	12	0.13
West Sak No. 9	2678-2930	35.5	42.2	50	0.20
West Sak No. B-10	2695-3174	37.6	41.9	55	0.11
West Sak No. 11	2324-2694	28.6	44.3	27	0.07
West Sak No. 17	3516-3620	15.8	31.3	12	0.12
West Sak No. 18	2526-2720	36.3	38.3	13	0.07
East Ugnu No. 1	3386-3654	32.6	44.2	27	0.10
Kuparuk No. 1G-7	3408-3638	24.8	34.6	64	0.28
Kuparuk No. 3B-14	2538-3025	38.3	41.0	136	0.28
Milne Pt. No. N-1B	3555-3806	24.2	24.1	56	0.22

* Represents interval that is potentially hydrocarbon bearing. Properties of individual West Sak sand members are listed in Tables 2-6.

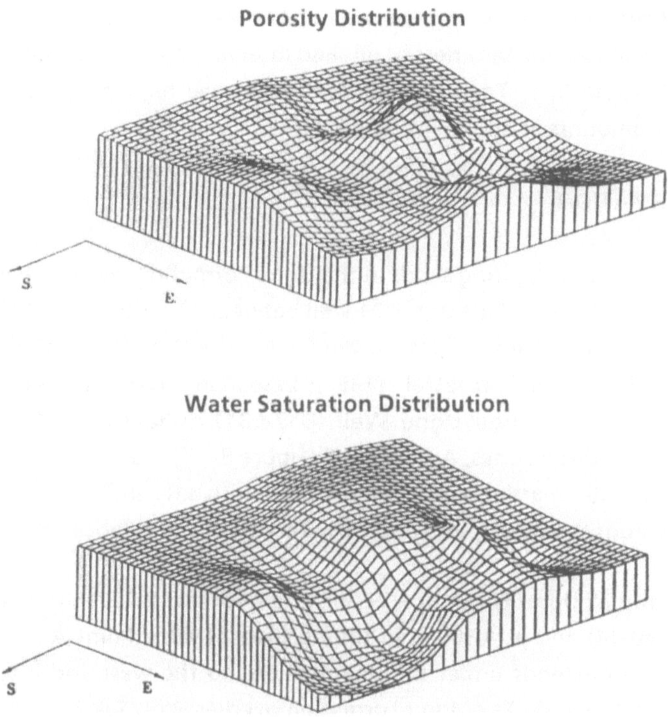

Figure 7. 3-D Plot Illustrating Porosity Distribution in West Sak Sands

sands deposited in the southwest are better sorted than those deposited in the northeast.

Reservoir Characteristics

The reservoir parameters, i.e. average porosities, water saturation, and net pay zone in the Upper and Lower sands units were computed from well logs and are listed in Tables II and III, and their spatial variations are shown in Figure 7. Subsequently, variations the hydrocarbon potentials of West Sak sands were evaluated first by determining the weighted means of effective porosity, water saturation and net pay thickness of all sand intervals in the West Sak Formation in each well (Table I). The sands of West Sak have high porosity and contain substantial amounts of oil.

UGNU FORMATION

The oil and tar-bearing sands in the Ugnu Formation were first encountered in the British Petroleum East Ugnu #1 well between the interval of 2716 and 3260 feet below surface. Werner[3] (1985) subdivided the Ugnu Formation into Upper and Lower Sands. Hallam et al[4] (1991), based on extensive analysis of core, recovered from dedicated Ugnu Well (DS2-0B1) drilled in 1989, divided the formation into three Zones, A through C (Figure 8). The Ugnu Formation consists of predominantly massive sands interbedded with silty shales deposited during Late Cretaceous (Maastrichtian) and Early Tertiary (Early Paleocene).

The top of Ugnu Formation occurs at a depth of 2,200 ft in the southwest (Kuparuk 3B-14) and at 4,300 ft in the northeast (Milne Point A-3) (Figure 10). The formation extends under the Beaufort Sea to the west and into the Milne Point Unit in the east. The Ugnu Formation overlies West Sak Formation and has similar regional structural trends. The beds dip northeasterly about 2° forming a monocline with northeast strike (Figure 9). There has been post-depositional faulting and regional tilting. Numerous north-south trending faults with mostly downthrow to the east traverse the Ugnu deposits.

It is estimated that Ugnu deposits contain over 15 billion barrels of oil-in-place. As a result of the proximity of the overlying permafrost, the reservoir temperature ranges from 45° to 100°F. Due to extensive biogradation, the viscosity of Ugnu oil is fairly high and varies between 50,000 and 10,000,000 cp at reservoir conditions. The gravity of oil varies from 8° to 14° API. According to Bryramjee[5] (1983), Ugnu crude can be classified as extra heavy oil to bitumen.

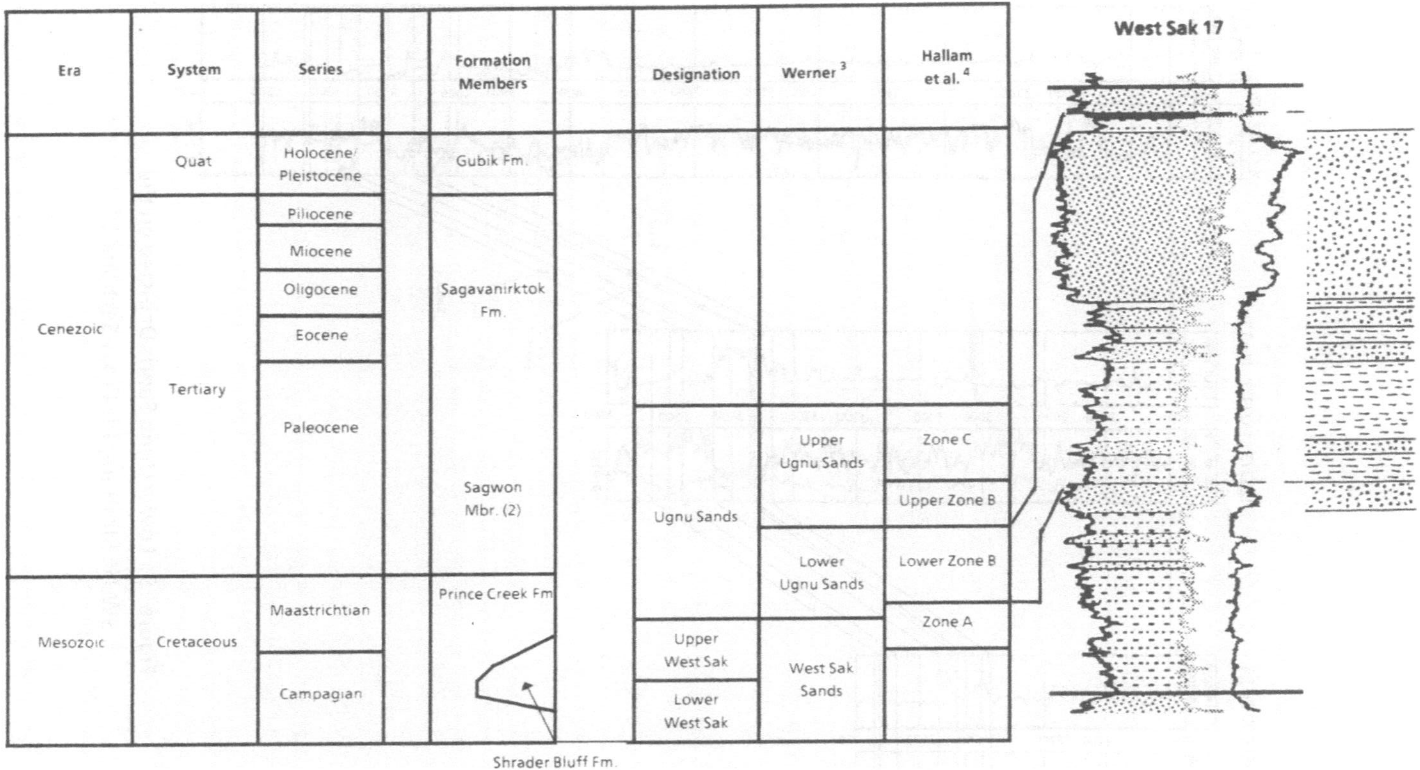

Figure 8. Generalized Stratigraphic Section of Upper Cretaceous and Lower Tertiary in West Sak and Ugnu Fields (modified from Werner[3] and Hallam[4])

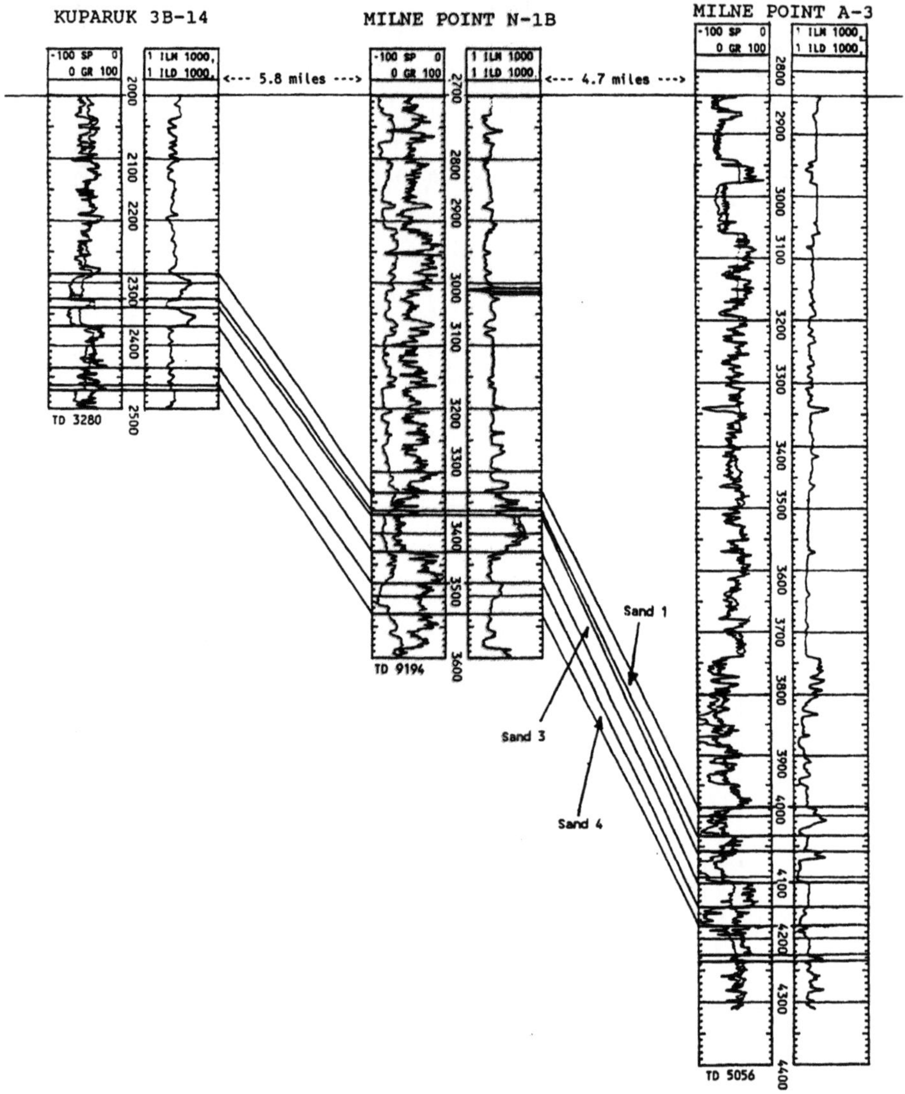

Figure 9. Lower Ugnu Sands Depicted on the SW-NE Stratigraphic Cross-Section CC '

Stratigraphy

The Ugnu Formation was deposited during the Late Masstrichtian of Cretaceous and Early Paleocene of Tertiary. The formation was subdivided into Upper and Lower Ugnu Sands by Werner[3] (1985). Detailed cores studied by Hallam et al.[4] (1991) revealed that Ugnu Formation consists of progradation of a lobate delta and beach ridge followed by transgression and partial destruction of the delta. They identified various deltaic and nearshore facies and subdivided Ugnu deposits into three zones, A (bottom) through C. The deltaic deposits from B Zone are further separated by a coal seam into Upper Zone B and Lower Zone B. The top unit of the Lower Zone A consists of massive sands of over 100 ft in thickness with good reservoir parameters, and contains by far the largest hydrocarbons of the Ugnu sands. This sand layer in the Lower Zone B corresponds to the Lower Ugnu Sand 1 described here.

The Zone C or the Lower Ugnu Formation does not contain hydrocarbons in significant quantity and therefore is not discussed. On the other hand, the top most sand layer in Zone C (Hallam, et al.[4] 1991) corresponds to the lower sand layer in the Lower Ugnu Formation (Werner[3], 1985) and is described here as Lower Ugnu Sand 5 (Figure 8).

The reservoir rock consists primarily of unconsolidated fine to medium grained, litharenites with quartz, rock fragments, feldspars and clay minerals. Sands are moderately to well sorted with good intergranular porosity. These immature sands are slightly compacted to uncompacted, and cores tend to explode after retrieval due to the trapped gas. The clay minerals which constitute 5% - 7% of the rock are kaolinite, chlorite and illite. Clays also occur as aggregates.

The sands of Ugnu Reservoir are interbedded with shales consisting of sandy silty clays. The clay minerals in shales predominantly consist of smectite and illite with kaolinite and chlorite. The percentage of silt in shale is approximately 45%. These interbedded shales form impervious seals and with varying thicknesses extend throughout the field (Figure 9).

Environment of Deposition

Detailed mineralogical and textural analysis of Ugnu Formation has been conducted by Mowatt et al.[6] (1991) and Hallam et al.[4] (1991). The sand bodies with good reservoir parameters in the Lower Ugnu Formation were deposited by the distributary channels of a prograding delta. The coarse channel sands are

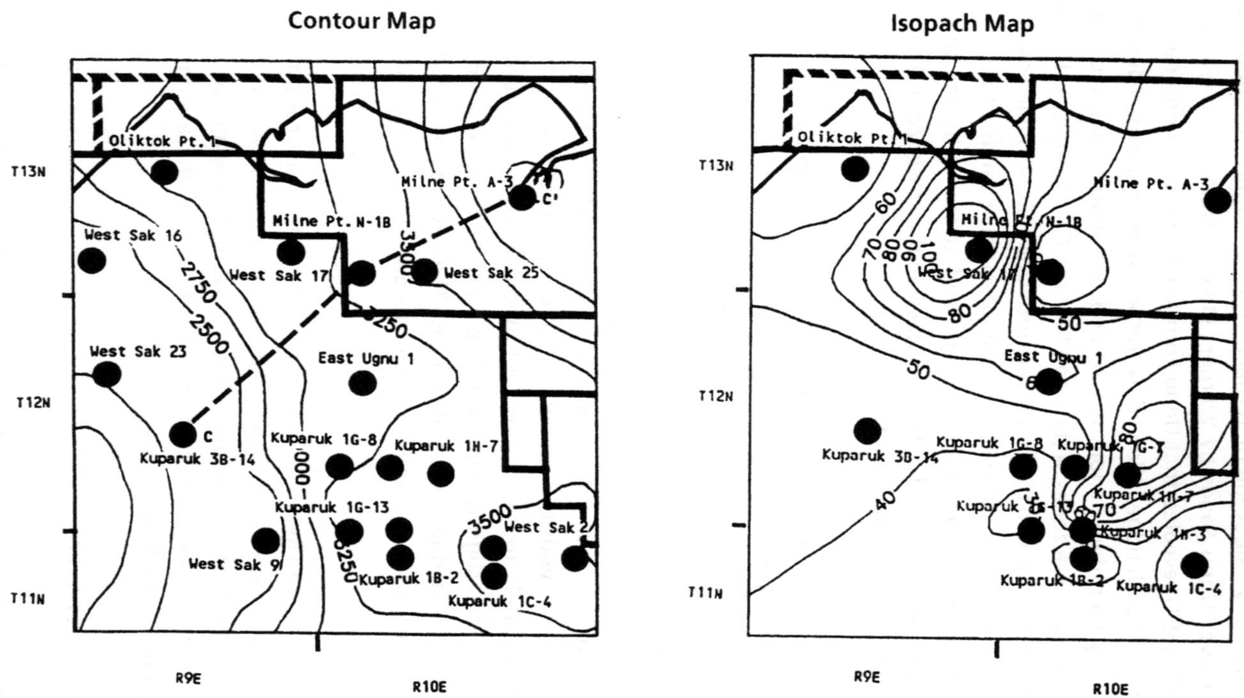

Figure 10. Isopach Map and Contour Map Showing Depth of the Top of the Lower Ugnu Sand 1

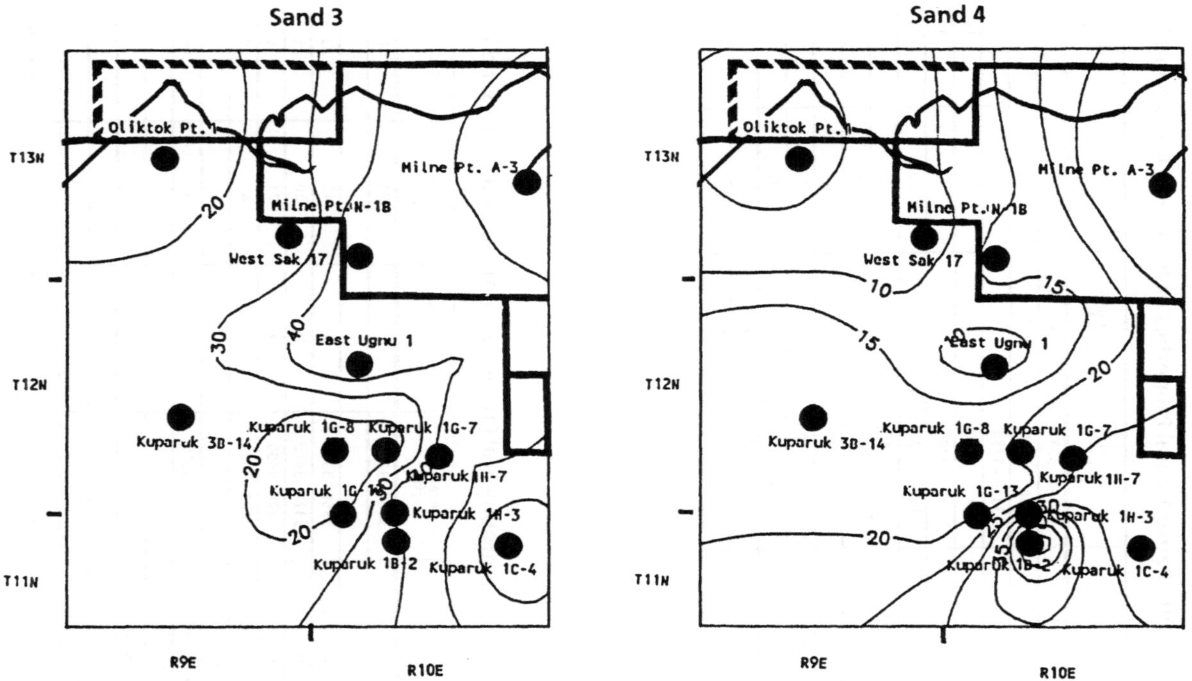

Figure 11. Isopach Maps Lower Ugnu Sands 3 and 4

Table IV. Petrophysical Properties of the Lower Ugnu Sand 1

Well Name	Interval (ft)	Average ϕ (%)	Average S_w (%)	Net Pay (ft)
West Sak No. 17	3122 - 3237	38	19	111
East Ugnu No. 1	3007 - 3060	35	22	52
Kuparuk No. 1B-2	3314 - 3338	38	40	16
Kuparuk No. 1H-3	3278 - 3370	26	14	75
Kuparuk No. 1H-7	3304 - 3386	38	19	49
Kuparuk No. 1G-7	3381 - 3428	40	22	24
Kuparuk No. 1G-8	3040 - 3080	23	23	6
Kuparuk No. 1G-13	3469 - 3505	35	46	19
Kuparuk No. 1C-4	3760 - 3799	28	27	21
Kuparuk No. 3B-14	2286 - 2328	42	23	43
Oliktok Pt. No. 1	3035 - 3096	27	36	3
Milne Pt. No. N-1B	3356 - 3389	31	15	28
Milne Pt. No. A-3	4266 - 4312	31	31	23

Table V. Petrophysical Properties of the Lower Ugnu Sand 2

Well Name	Interval (ft)	Average ϕ (%)	Average S_w (%)	Net Pay (ft)
West Sak No. 17	3266 - 3280	16	74	0
East Ugnu No. 1	-	-	-	-
Kuparuk No. 1B-2	3356 - 3388	40	40	23
Kuparuk No. 1H-3	3378 - 3410	36	20	25
Kuparuk No. 1H-7	3470 - 3490	16	71	0
Kuparuk No. 1G-7	3465 - 3514	36	42	29
Kuparuk No. 1G-8	-	-	-	-
Kuparuk No. 1G-13	3535 - 3619	36	47	46
Kuparuk No. 1C-4	-	-	-	-
Kuparuk No. 3B-14	-	-	-	-
Oliktok Pt. No. 1	-	-	-	-
Milne Pt. No. N-1B	-	-	-	-
Milne Pt. No. A-3	-	-	-	-

flanked by levies and lateral facies of finer sand interdistributary crevasse sand, silt and shaly silts. Cleaner sands deposited at the mouth of these distributary channels as mouth bar are also common. Thick sand layers observed in the Lower Ugnu Formation are due to the stacking of channels during seaward progradation of the delta. Therefore, reservoir sands do not form a continuous sand sheet, rather interconnected cylindrical elongated sand deposits.

The presence of coal seams suggests that there were periods of abandonment of delta due to the lowering of sea level. Because the reservoir sands are restricted to channels and their subsequent subaerial exposure and erosion, it is difficult to precisely correlate these sands between wells. In particular, the thickness of these sand bodies changes abruptly and some of the sand layers are absent in some wells. However, the major sand layers are easily identified throughout the field. It should be noted that channel sands observed in one well may not laterally extend into the adjacent well and may correlate to the sands representing other channels on the delta. These channel sands characteristically contain higher contents of fines near the bottom and become cleaner and coarser near the top.

The maximum thickness of these channel sands occurs in the Milne Point N-19 and tends to taper towards south and southwest (Figure 9) It appears that these sands were deposited on an east-west oriented delta. The thickness of each sand unit varies considerably as illustrated by the isopach maps in Figures 10 and 11.

In the Lower Ugnu Formation, Sand 1 forms the thickest reservoir and is distinctly present in all the wells studied (Table IV). Sand 3 and Sand 4 extend laterally with varying thickness (Tables VI and VII). On the other hand, Sand 2 and Sand 5 were observed in a few wells only (Tables V and VIII). It appears that these sand units were probably truncated due to subaerial erosion. It should be noted that while the thickness of Sand 3 increases toward east and northeast, Sand 4 increases in thickness toward the southeast with a depocenter in the vicinity of Kuparuk 1B-2.

Reservoir Characteristics

The average porosities and water saturations of sand units in the Lower Ugnu Formation are shown in Tables IV and VIII. The porosity in Sand 1 is quite high and varies from 23% to 40% (Table IV), and water saturation ranges between 14% and 45%. On the average water saturation is below 25%. The net pay thickness in Sand 1 exceeds 20 ft in most wells and is over 110 ft in West Sak 17.

Table VI. Petrophysical Properties of the Lower Ugnu Sand 3

Well Name	Interval (ft)	Average ϕ (%)	Average S_w (%)	Net Pay (ft)
West Sak No. 17	3368 - 3390	38	34	14
East Ugnu No. 1	3130 - 3180	33	26	45
Kuparuk No. 1B-2	3479 - 3531	33	40	32
Kuparuk No. 1H-3	3456 - 3510	38	43	45
Kuparuk No. 1H-7	3482 - 3486	21	62	0
Kuparuk No. 1G-7	3610 - 3620	25	34	4
Kuparuk No. 1G-8	3126 - 3140	30	4	7
Kuparuk No. 1G-13	3707 - 3732	26	48	2
Kuparuk No. 1C-4	3826 - 3902	28	65	0
Kuparuk No. 3B-14	2340 - 2370	38	35	30
Oliktok Pt. No. 1	3114 - 3137	36	94	0
Milne Pt. No. N-1B	3396 - 3464	38	9	62
Milne Pt. No. A-3	4336 - 4388	30	32	9

Table VII. Petrophysical Properties of the Lower Ugnu Sand 4

Well Name	Interval (ft)	Average ϕ (%)	Average S_w (%)	Net Pay (ft)
West Sak No. 17	3404 - 3412	25	45	2
East Ugnu No. 1	3233 - 3292	33	44	46
Kuparuk No. 1B-2	3560 - 3632	34	42	42
Kuparuk No. 1H-3	3568 - 3592	39	47	14
Kuparuk No. 1H-7	3518 - 3564	37	54	0
Kuparuk No. 1G-7	3715 - 3748	38	45	17
Kuparuk No. 1G-8	3194 - 3212	26	15	10
Kuparuk No. 1G-13	3834 - 3854	39	48	3
Kuparuk No. 1C-4	3988 - 4023	9	71	0
Kuparuk No. 3B-14	2436 - 2464	38	41	29
Oliktok Pt. No. 1	3174 - 3193	26	97	0
Milne Pt. No. N-1B	3514 - 3570	18	30	20
Milne Pt. No. A-3	4430 - 4460	11	100	0

Table VIII. Petrophysical Properties of the Lower Ugnu Sand 5

Well Name	Interval (ft)	Average ϕ (%)	Average S_w (%)	Net Pay (ft)
West Sak No. 17	3422 - 3462	25	74	0
East Ugnu No. 1	-	-	-	-
Kuparuk No. 1B-2	3707 - 3728	30	39	12
Kuparuk No. 1H-3	3610 - 3630	41	50	1
Kuparuk No. 1H-7	3636 - 3662	36	59	0
Kuparuk No. 1G-7	3755 - 3790	32	48	4
Kuparuk No. 1G-8	-	-	-	-
Kuparuk No. 1G-13	3878 - 3901	34	47	2
Kuparuk No. 1C-4	-	-	-	-
Kuparuk No. 3B-14	2470 - 2490	41	34	20
Oliktok Pt. No. 1	3309 - 3318	24	100	0
Milne Pt. No. N-1B	-	-	-	-
Milne Pt. No. A-3	4504 - 4514	15	25	3

Table IX. Log Derived Data for Selected Wells in the Lower Ugnu Sands

Well Name	Interval (ft) (Measured Depth)	Weighted Mean Effective Porosity (%)	Weighted Mean Water Saturation (%)	Net Pay Thickness (ft)	Hydrocarbon Pore Volume (10^3bbl/acre-ft)
West Sak No. 17	3122 - 3462	37	21	127	294.3
East Ugnu No. 1	3007 - 3292	34	30	142	260.4
Kuparuk No. 1B-2	3314 - 3728	35	41	128	206.5
Kuparuk No. 1H-3	3278 - 3630	32	26	161	289.1
Kuparuk No. 1H-7	3304 - 3662	38	19	49	115.4
Kuparuk No. 1G-7	3381 - 3790	37	36	77	384.5
Kuparuk No. 1G-8	3040 - 3212	26	14	22	38.1
Kuparuk No. 1G-13	3469 - 3901	36	46	71	354.6
Kuparuk No. 1C-4	3760 - 4023	28	27	21	33.0
Kuparuk No. 3B-14	2286 - 2490	40	32	121	255.3
Oliktok Pt. No. 1	3035 - 3318	27	36	3	4.3
Milne Pt. No. N-1B	3356 - 3570	32	14	110	236.0
Milne Pt. No. A-3	4266 - 4514	30	31	35	55.2

Contour Map

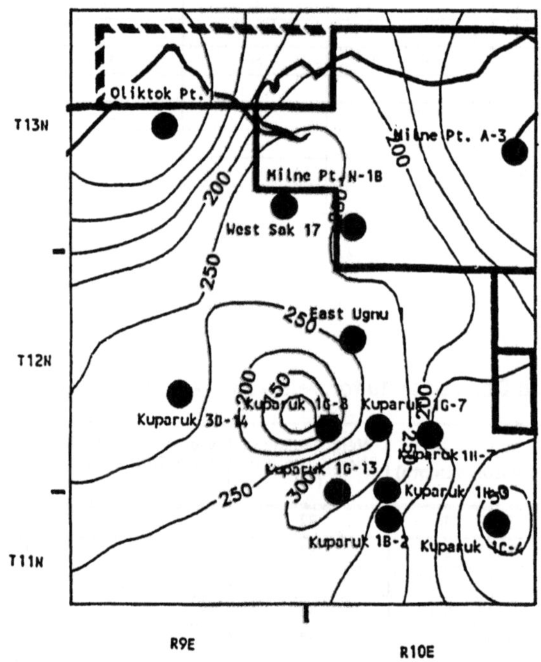

Figure 12. Contour Map of Hydrocarbon Pore Volume (q0³ bbl/acre - ft)

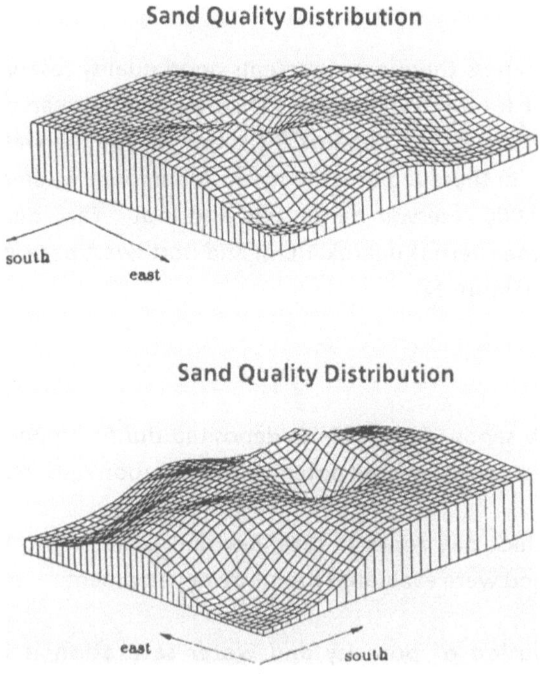

Figure 12. Sand Quality Distribution in the Lower Ugnu Sands

Sand 3 and Sand 4 also have good porosities (Tables VI and VII). The water saturation in these sands, however, is also high and ranges from 30% to 50% in most wells. Sand 2 and Sand 5, observed in only a few wells, also have good porosity but have high water saturation. These sand units are not widely distributed and thus were not studied in detail.

The weighted average reservoir parameters of the Lower Ugnu Sands are listed in Table IX. The average porosity and the average water saturation in these sands varies from 27% to 28% and from 14% to 46% respectively. Net pay zone thickness ranges from 3 ft to 161 ft. In most wells, the net pay thickness exceeds 50 ft.

The net sand thickness therefore represents good quality reservoirs which could be considered for future development. The best quality reservoir sand occurs in the northeastern part of the Kuparuk River Unit and in the western part of the Milne Point Unit. In these regions, the hydrocarbon pore volume varies between 200,000 and 250,000 reservoir barrels per acre foot. The quality of reservoir sands in the southeastern (Kuparuk 1 C-4) and northwest parts (Oliktok Point 1), however, is poor (Figure 12).

CONCLUSIONS

1. The shallow sands of large delta deposited during Upper Cretaceous and Lower Tertiary contain large potential hydrocarbon resources.

2. Relatively thick pay zones in West Sak and Lower Ugnu formations were identified and were evaluated for future development.

3. The distribution of porosity and water saturation is complex and is controlled primarily by the depositional environment and post-deposition burial.

4. These reservoir sands are semi-consolidated to unconsolidated and therefore, in some regions will require special well completion and production techniques.

5. Commercial production from the shallow parts of the reservoirs with low temperatures and high viscosity oil will require use of thermal and miscible enhanced oil recovery techniques.

6. Most oil occurs in the narrow channel deposits which are vertically stacked. Recovery of oil through displacement by water flooding and thermal injection will require infill drilling. Therefore knowledge of the reservoir framework will be the key to the development of these shallow sands.

BIBLIOGRAPHY

1. "LOGCALC: Users Manual", Scientific Software Intercomp, Denver, CO, (1986)

2. Lanning, E.N. and Johnson, D.M., "Automated Identification of Rock Boundaries: An Application of the Walsh Transform to Geophysical Well-log Analysis", Geophysics, Vol. 48, No. 2, (1983)

3. Werner, M.R., "West Sak and Ugnu Sands - Low Gravity Oil Zones of Kuparuk Area, Alaskan North Slope," AAPG Bull. Vol. 23, (1985)

4. Hallam, R.J., Piekenbrock, E.J., Abou-Sayed, A.S., Garon, A.M., Putnam, T.W., Weggeland, M.C. and Webb, K.J.," Resource Description and Development Issues for the Ugnu Reservoir, North Slope, Alaska", SPE 21779, presented at the Society of Petroleum Engineers Western Regional Meeting, Long Beach, CA, March 20-22, (1991)

5. Byramjee, R.J., "Heavy Crudes and Bitumen Categorized to Help Assess Resources, Techniques," OGJ, Vol. 81, No. 27, pp 78-82, (1983)

6. Mowatt, T.C., Ogbe, D.O., Kamath, V.A., and Sharma, G.D., "Petrologic - Petrophysical Relationships, West Sak and Ugnu (Brookian), Northern Alaska," AAPG Bull. Vol. 75. No. 3, (1991)

THE POTENTIAL OF NATURAL GAS IN ALASKAN ARCTIC

G. D. Sharma, V. A. Kamath and S. L. Patil

Petroleum Development Laboratory
425 Duckering Building
University of Alaska Fairbanks
Fairbanks, AK 99775

The proven reserves of natural gas in northern Alaska are estimated to be 47 trillion standard cubic feet (TSCF) and represent approximately 25% of the total U.S. reserves. The undiscovered recoverable conventional natural gas resources are estimated to 89 TSCF and additional 500 TSCF are deposited as unconventional gas hydrates. The hydrates are mostly found along the North Slope and adjacent shelf.

Presently, abundantly available natural gas on the North Slope has no market and most of the produced gas is reinjected. Continued reinjection will ultimately result in high gas/oil ratio in the reservoir thus causing gas recycling problem and production decline. The purpose of this paper is to review the available conventional and unconventional natural gas resources of northern Alaska and their potential uses.

The various options for utilization of natural gas from northern Alaska include: (i) Lower 48 States market, (ii) Pacific-Rim markets, (iii) conversion to fuel grade methanol and/or gasoline, (iv) use of natural gas for EOR processes, (v) gas based petrochemicals and, (vi) gas-utilities systems. Among all the options discussed it appears that most expedient use of the North Slope gas is for the enhanced oil recovery. Conversion of the natural gas into methanol and its transport through existing pipeline as well as transportation of natural gas in the form of liquified natural gas to Pacific Rim countries could also become economically viable options.

INTRODUCTION

The natural gas deposits on the North Slope, Alaska is a significant resource compared to the one in the continental United States. Currently, the proven reserves of natural gas in Alaska are estimated to be 47 TSCF compared to 140.0 TSCF in the lower 48 states.

Particle Technology and Surface Phenomena in Minerals and Petroleum
Edited by M.K. Sharma and G.D. Sharma, Plenum Press, New York, 1991

The U.S. can not be a market for North Slope gas even though it is a net importer of natural gas. Indigenous production in Lower 48 States combined with imports from inexpensive natrual gas sources in Canada, Mexico and Africa are expected to meet projected demands for natural gas for the next few decades. On the other hand, the Pacific Rim countries such as Japan, Korea and Taiwan might import the liquified natural gas from North Slope, Alaska as a reliable source if the price is not significantly different from the worldwide prices.

In this paper the following areas regarding the natural gas on North Slope, Alaska are outlined.

1. The geology of North Slope, Alaska in reference to the oil and gas accumulations.
2. Extent of natural gas reserves on the North Slope, Alaska.
3. Alternatives for utilization of natural gas in the Arctic.

GEOLOGICAL SETTINGS OF NORTH SLOPE GAS FIELDS

The North Slope region consists of three major packages of rocks deposited in sequences. The oldest package of rocks was deposited prior to Devonian period in an east west oriented Colville Geosyncline. These rocks were subsequently metamorphosed and now called basement rocks. During Devonian time the metamorphosed rocks were uplifted to form a broad stable Arctic Platform which extended along northern Alaska. The shallow marine platform initially received carbonate sediments which were proceeded by sand and shales. Marine sedimentation continued from Late Devonian or Early Mississippian through Early Cretaceous time and formed the second package of rocks. At the beginning of the Cretaceous Period, intense tectonic activity compressed the southern part of the platform and thrusted Paleozoic to Cretaceous rocks northward onto the central part of the platform. The large scale compression resulted in the formation of the present Brooks Range - Hearld Arch which subsequently supplied sediments to the north for the third package. These consists of rocks ranging from Early Cretaceous to Middle Eocene which provided with the major filling of the basin. During the deposition of the third package, the region was intermittently uplifted and thus the area was exposed and submerged under the sea repeatedly.

From the Middle Eocene until Middle Pliocene time the North Slope region was above sea level. The region may have submerged briefly and sedimentation may have occurred in some areas. During this period a stream-sculptured topography similar to that observed today probably existed on the North Slope region.

During Pleistocene Epoch, the crustal warping resulted in marine transgressions and regressions. The sediments deposited during Pleistocene covered the underlying Tertiary topography and formed a monotonously flat platform.

The North Slope region is subdivided tectonically into three major east-west trending structures. (Figure 1). These are the Barrow Arch, Colville Trough, and Brooks Range. The Barrow Arch is a subsurface broad regional structure. The axis of this arch follows the Beaufort Sea Coast from the foothills of the Brooks Range in the east to Point Barrow and continues under the Chukchi Sea in the west. The arch was developed during Late Jurassic or Early Cretaceous time on the northern province of the ancient Arctic Platform.

The east-west oriented Colville Geosyncline was a major broad depression during Paleozoic and Mesozoic eras and subsequently included the Barrow Arch in the north and the Brooks Range in the south. It is filled with deposits ranging from early Paleozoic to recent in age. The sediments deposited in this basin, excluding Paleozoic which have been metamorphosed, are in excess of 30,000 ft.

The Brooks Range is dominated by major thrust faults which produced complex series of thrust belts and sheets. Northern parts of the range, the Foothill Fold Belt, consists of shallow folds with complex axial shale cores while southern zone comprises of complex folds and thrust faults. Several structures in the northern part of the Foothill Fold Belt contain hydrocarbons.

A generalized stratigraphic sequence of the North Slope region is provided in Figure 2. Based on investigations by Jones and Speers[1] (1976), Young et al[2] (1977), Seifert et al[3] (1979), and Magoon and Claypool[4] (1981) the sources of oil and gas in North Slope are the Triassic Shublik Formation, a phosphatic-calcareous shale; the Jurassic Kingak Shale (average age of oil 220 m.y.) and deeply buried Upper Cretaceous shales (post-Neocomian Shale, average age of oil 87 m.y.). It therefore appears that the second package and the third package of sediments deposited in the Colville Geosyncline had sufficient source sediments to provide abundant hydrocarbons.

The hydrocarbons evolving from the Paleozoic package have been trapped in the super giant Prudhoe Bay Field and after filling the trap spilled westward into Kuparuk River, West Sak and Ugnu fields. The combined oil in place is estimated to be 65 billion barrels. Prudhoe Bay complex also contains about 38 TCF of natural gas. The westward migration of oil also resulted in trapping of oil in younger (shallower) traps and thus causing increasing degradation of oil and loss of gas and lighter components. Part of the natural gas escaping from Super

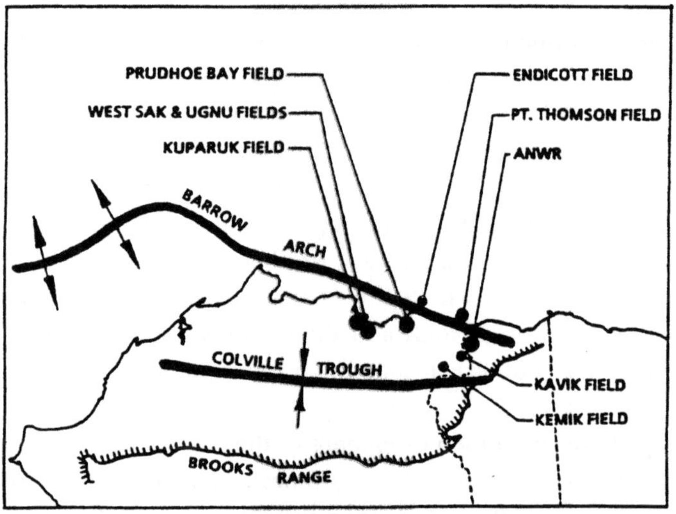

Figure 1. Alaskan North Slope Basin Showing Barrow Arch, Colville Trough, and Brooks Range

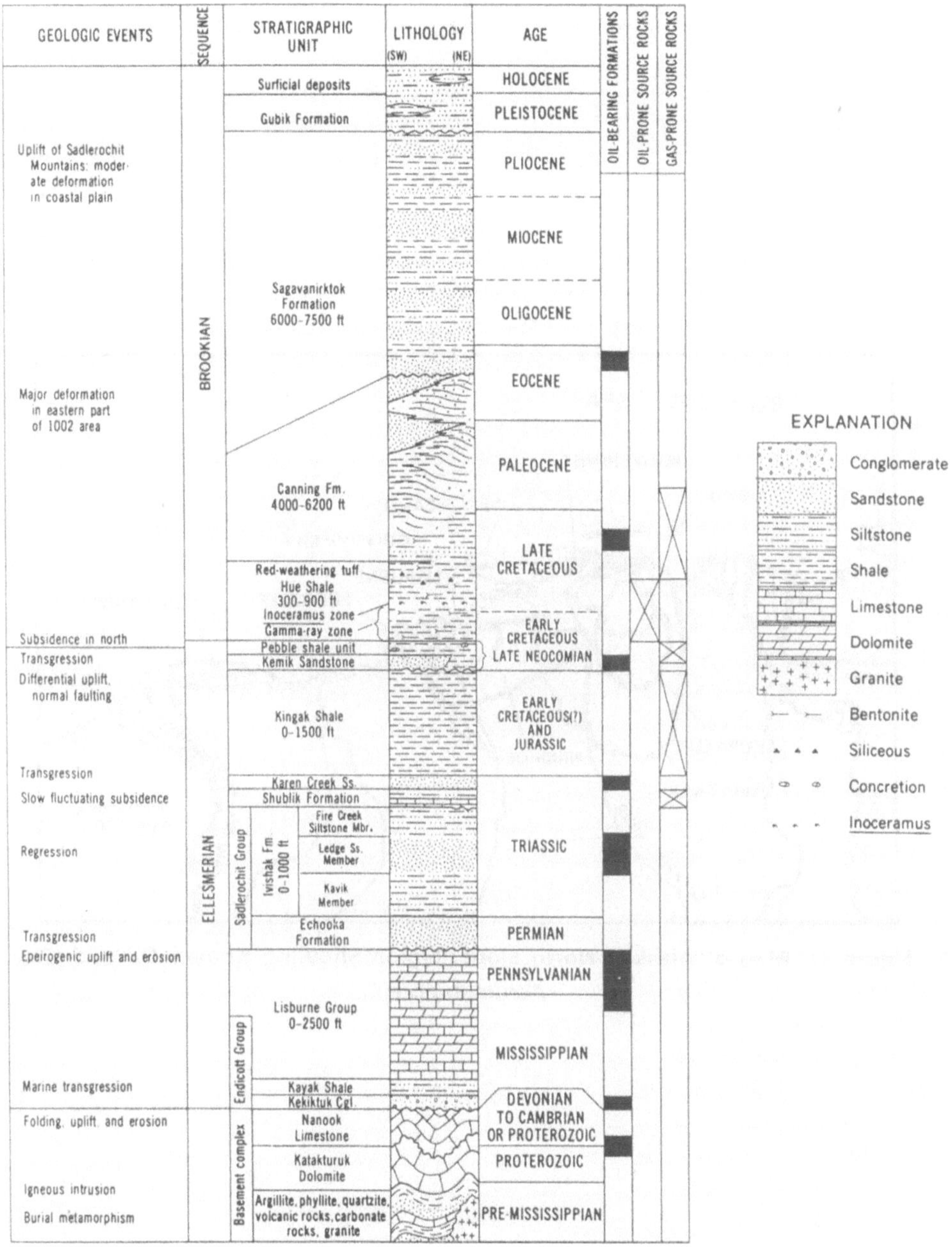

Figure 2. Generalized Stratigraphic Column Showing Occurrence of Petroleum on the Alaskan North Slope Region (Source: Explorer - AAPG; June 1987)

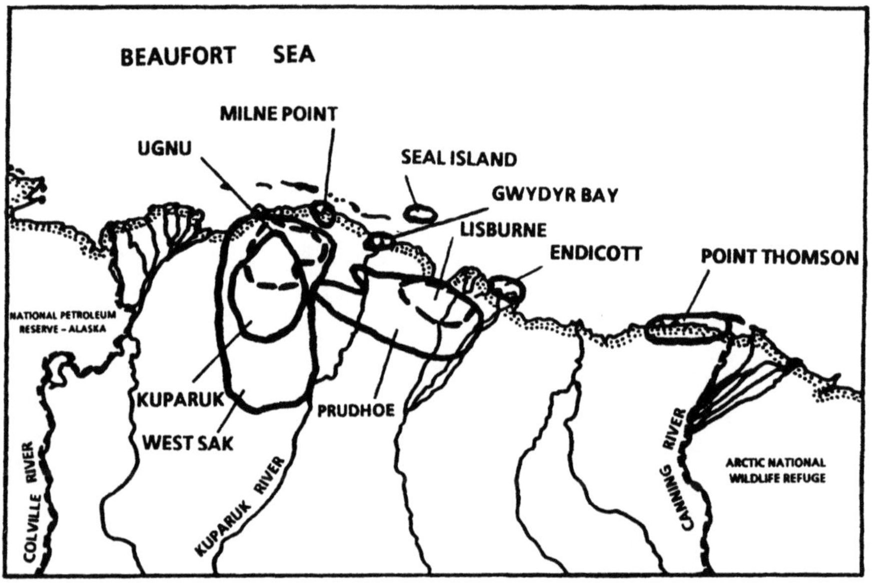

Figure 3. Map of Alaskan North Slope Region Showing Known Oil and Gas Accumulations

Giant West Sak and Ugnu sands is trapped under the permafrost as gas hydrates. The amount of natural gas trapped as gas hydrate in northern Alaska is estimated at 500 TCF (Lewin and Associates[5], 1983). The westward migration, on the other hand resulted in entrapment of hydrocarbons in older rocks such as Lisburne Formation (Mississippian age).

The Late Cretaceous hydrocarbons from the third package have been found in various reservoirs along the northwest trending region between Umiat and Cape Simpson in the National Petroleum Reserve in Alaska, and along the east-west trending Barrow Arch. The presence of Cretaceous oil and gas has been observed as far east as Ungoon Point (near U.S. - Canada border). An eastern migration of Cretaceous oil and gas has been inferred from drilling of Mukluk Well and the surrounding area.

The potential reservoir beds in the North Slope region of Pre-Tertiary age in ascending order are:

1. Mississippian - Pennsylvanian clastic facies of the Lisburne Group;
2. Mississippian "Dolomite Unit" of the Alahap Limestone;
3. Pennsylvanian - Permian Wahoo Limestone;
4. Permian Ikiakpaurak member of the Echooka Formation;
5. Triassic "Ivishak Sandstone";
6. Triassic Sag River Sandstone;
7. Early Jurassic "Barrow Sandstone";
8. Late Jurassic "Simpson Sandstone";
9. Jurassic - Neocomian Kingak Formation; Includes potential reservoirs "Peard Sandstones", Neocomian "Kealok Anomaly"; and
10. Cretaceous "Pebble Shale" sandstone, "Walapka Sandstone", "Basal Torok" and Marine and Non-marine sandstones of the Nanushuk Group.

The known oil and gas reservoirs on the North Slope fields are shown in Figure 3. Oil and gas occurs over a wide area at various horizons. Source rock was abundant in Paleozoic and Upper Cretaceous/Tertiary to generate sufficient quantity of hydrocarbons to fill traps of various resrvoirs. Gas from Paleozoic Package occurs in Triassic Sadlerochit Formation near Barrow in the east, in Kemik and Kavik in the west, and in Seal Island, and Gwydyr Bay in the north. Lower Cretaceous gas have been observed in Prudhoe Bay complex, Kuparuk, Milne Point, Point Thomson and the Beaufort-Mackenzie areas, Canada in the east.

The associated gas from Upper Cretaceous/Tertiary package has been observed in Umiat and Gubik in the south, Simpson in the west and Mackenzie Delta, Canada in the east.

From the distribution of hydrocarbons on the North Slope (the area between the National Petroleum Reserve Alaska [NPR-A] and the Arctic National Wildlife Refuge [ANWR] approximately 65,000 sq. mi.) it is apparent that the source rock during catagenesis produced abundant crude as well as natural gas. Due to abundance of produced hydrocarbons, these could not be accommodated in a single reservoir and migrated laterally and vertically thus producing many super giant fields. Prudhoe Bay Field being closer to the source was probably the first trap to be filled. Further migration of hydrocarbons to the west filled traps in Kuparuk River, West Sak and Ugnu reservoirs while free, natural gas formed hydrates and other gas fields to the west.

Natural gas from Paleozoic and Tertiary-Cretaceous these packages is distributed even to a larger area. Gas from packages have been observed throughout the North Slope. The distribution of oil and natural gas originating from the second package lead to believe that liquid hydrocarbons migrated to a lesser distance and occupied traps in the vacinity of the Prudhoe Bay and the regions to the west. On the other hand gas migrated to greater extent and has been observed as far as Barrow in the west and in Kemik and Kavik field in the east. Therefore, it is conceivable that not all the natural gas which potentially occurs in this region has been discovered. Lack of discovery may be due to the lack of market for the North Slope gas and industry is not actively exploring for gas. This is supported by the shut-in status of major gas fields in northern Alaska.

EXTENT OF ALASKAN NATURAL GAS RESERVES

A full evaluation of natural gas resources in northern Alaska must also include estimates of potential gas in both onshore and offshore regions where potential subsurface gas reservoirs remain unexplored.

Commercial natural gas occurrence can be broadly classified into two categories, conventional and unconventional. The conventional category includes natural gas found as associated and non-associated or free gases, including dissolved or solution. The unconventional natural gas resource includes, (1) deep gas, (2) coal bed methane, (3) gas hydrates and (4) gas in shales, (5)

TABLE I

NORTHERN ALASKA DISCOVERED NATURAL GAS RESOURCES
(In Trillion SCF)

	DESIGNATED OPERATOR	PROVEN RESERVES	PRODUCING FORMATION
I. ASSOCIATED			
Prudhoe Bay Field	ARCO/Standard Alaska	38.26	Sadlerochit
Kuparuk River Field	ARCO	0.56	Kuparuk River
Endicott Field (Duck Is. Unit)	Standard Alaska/ Exxon	0.80	Kekituk
Lisburne Field	ARCO	1.10	Lisburne
Milne Point Field	Conoco	0.25	Kuparuk River
Point Thomson	Exxon	6.00	Thomson Sands
Colville Delta	Texaco	Not Available	
North Star	Amerada Hess	Not Available	
Niakuk Area	Standard	Not Available	Sag River/ Sadlerochit
Gwydyr Bay	Conoco	Not Available	Sadlerochit
Umiat		Not Available	
II. NON-ASSOCIATED GAS			
East Barrow		0.01	Jurassic
South Barrow Field		0.01	
Point Thomson		Shut-in	
Kemik	Forest Oil Corp.	Shut-in	Shublik Formation
Kavik	Amoco	Shut-in	Sag River/Sadlerochit
East Umiat	BP Expl. Co. (Alaska) Inc.	Shut-in	Ninuluk/Chandler
TOTAL		46.99	

Sources: Alaska Department of Natural Resources 1986 and Alaska Oil and Gas Conservation Commission 1987.

TABLE II

NORTHERN ALASKAN UNDISCOVERED CONVENTIONAL NATURAL GAS RESOURCES
(In Trillion SCF)

ONSHORE	
Arctic National Wildlife Refuge (ANWR)	31.3
National Petroleum Reserve - Alaska (NPRA)	14.6
Unexplored North Slope Regions	12.9
Subtotal	58.8
OFFSHORE	
Beaufort Shelf (<60m water depth)	5.62
Beaufort Slope	6.70
N. Chukchi Shelf (<60m water depth)	15.1
S. Chukchi Shelf - Hope Basin (<60m water depth)	1.1
N. Chukchi Slope	1.7
Subtotal	30.22
TOTAL	89.02

geopressurized aquifers and tight sands. With the exception of gas hydrates, very little information for unconventional natural gas resources in Alaska is available.

Most oil fields discovered in northern Alaska have associated gas. While all have dissolved gas some have large gas caps. The production histories of these fields have now provided the amount of natural gas available and thus can be labeled as proven reserves. The current production of dissolved natural gas from some of these fields, in processed or unprocessed form, is reinjected into the field.

Table I presents a summary of the discovered natural gas resources in northern Alaska. It can be seen that the 38 trillion SCF natural gas in the Prudhoe Bay field is the most attractive natural gas resource and constitutes 81% of the total proven reserves. The other associated gas proven reserves are from Kuparuk River, Endicott, Lisburne, Milne Point and Point Thomson fields.

The northern Alaskan undiscovered conventional natural gas resource estimates are given in Table II. It should be noted that the 31.3 trillion SCF of estimated gas in the Arctic National Wildlife Refuge (ANWR) corresponds to an approximately 10% probability.

The natural gas hydrates are expected to contain more than 500 trillion SCF natural gas (Lewin and Associates[5], 1983). Gas hydrates are solid, crystalline, ice-like compounds composed of natural gas and ice. Gas hydrates have very high natural gas content, typically 170-180 SCF per cubic feet of hydrates. The production of natural gas from hydrates necessitates the dissociation of solid hydrates into natural gas and water. It takes about 10% of the heat content of produced natural gas to dissociate the gas hydrates. Godbole et al.[6] (1988) and Roadifer et al.[7] (1987) have presented nomograms to calculate the zone of hydrate stability onshore and offshore Alaska respectively.

Table III represents the northern Alaska natural gas resource summary for conventional as well as unconventional resources. It should be noted that the undiscovered unconventional resource of natural gas in the form of gas hydrates is only a rough estimate.

ALTERNATIVES FOR UTILIZATION OF NATURAL GAS OF THE ALASKAN ARCTIC

While the natural gas in the Alaskan Arctic represents a truly significant resource; a resource which should be exploited in the most appropriate manner, it

TABLE III

NORTHERN ALASKA NATURAL GAS RESOURCE SUMMARY

(In Trillion SCF)

I. CONVENTIONAL	PROVEN RESERVES	PROBABLE RESERVES	UNDISCOVERED RESOURCES
Onshore	46.19	7.80	58.80
Offshore	0.8	---	30.22
	46.99	7.80	89.02

II. UNCONVENTIONAL	PROVEN RESERVES	PROBABLE RESERVES	UNDISCOVERED RESOURCES
Gas Hydrates	---	---	500.0
TOTAL	46.99	7.80	589.02

is still not being utilized in any significant way. Only a small fraction of the gas is being used to operate some of the oil pumping stations for TAPS and for North Slope oil-field operations. Part of the produced gas is processed and is reinjected for miscible displacement of oilfrom the Prudhoe Bay Field. Most of the produced gas, however, is being reinjected as is required by the government regulations. For more than a decade, various options of utilization of Alaskan natural gas have been considered, examined, discussed and rediscussed (Chem Systems Inc.[8], 1979; Economides and Ostermann[9], 1982; Corley and Marsden[10], 1984). Wide range of viewpoints currently exist among various bodies such as State and Federal governments, pipeline companies, oil and gas industries and other private and public sectors. In 1987, U.S. DOE, State of Alaska and University of Alaska, Fairbanks co-sponsored a workshop to re-examine the various options for utilization of natural gas in the arctic. This section reviews various options with respect to merits and demerits addressing engineering, economic, environmental, supply/demand, market and political aspects and current situation.

<u>Option 1: Transportation of Natural Gas through ANGTS pipeline to U.S. Markets</u>

In 1976, Canada's Alcan and Salt Lake City's Northwest Pipeline Corporation submitted a proposal to build a 4,800 mile long, 48 inch pipeline called as Alaska Natural Gas Transportation System (ANGTS). The pipeline would deliver 2 BCF/D of natural gas with expansion capability of 3.2 BCF/D (Wadlington", 1987). This pipeline would run parallel to the existing TAPS from the North Slope to Delta Junction, then follow the Alaska Highway in Canada and South to the U.S. Border. At the borderline would split into an eastern leg and a western leg. The western leg would carry 30% of the gas to south through Washington and Oregon to a termination point near San Francisco. The eastern leg would carry 70% of the gas to its terminus, south of Chicago. In November 1977, this proposal was accepted by both the U.S. and Canadian governments to provide low cost fuel and to eliviate the perceived shortages of natural gas in U.S. and Canada at that time. In 1978, the estimated cost of the project was $10 billion in addition to the necessary $4.4 billion gas conditioning plant at Prudhoe Bay (Daily News Miner[12], Nov. 18, 1987).

As a result of Canadian Northern Pipeline Act of 1978 pursuant to the 1977 Transit Pipeline Treaty between U.S. and Canada, and laws passed by the U.S. Congress in 1981, Foothills Pipe Lines Ltd. was authorized and prebuilt 1,500 miles of the Canadian segment of the ANGTS and placed facilities in service during 1981 and 1982 to carry Canadian gas to the Midwest and West Coast of the U.S. (Jesperson[13], 1987) The construction of ANGTS, however, was delayed due to problems of economics and financing. Other problems emerged as well. The additional gas discoveries in the overthrust belt of the Rocky Mountains, Alberta, Canada and in Powder River Basin, Wyoming, increased the availability of less-expensive natural gas. The Northwest Alaskan Pipeline Co.'s latest official cost estimates of 1982, were $23 billion for the entire project. Many believed that this estimate was too low. The operating costs were estimated to be $170 million per year (Economides and Ostermann[9], 1982) and real costs for construction were estimated at $40 billion.

As of today, the project is delayed indefinitely due to two major problems. One is that the partner's have committed no funds for this project. ARCO Alaska, Inc. in December 1987 had pulled out of the financing deal (Daily News Miner[14], Dec. 14, 1987). Secondly, there is no market for the Alaskan gas in the lower 48 states due to high cost of transportation. The 1982 study, (Economides and Ostermann[9], 1982) shows delivery price of natural gas at $18 per MCF. Though at one time, this project looked like the most attractive option and though some still believe that this is the best option and ANGTS pipeline still may be built, however, under current supply and demand situations, this option is certainly not economical. Other options look much more attractive when compared to this option.

Option 2: Conversion of Natural Gas to Liquified Natural Gas andTransporation to Pacific Rim Markets

Yukon Pacific Corporation (YPC) believes that a significant market for Alaskan natural gas will open up in mid-1990's in the Pacific Rim countries like Japan, South Korea and Taiwan and help reduce the big trade deficits of U.S. (Treadwell[15], 1987). In 1984, YPC proposed to develop Trans-Alaska Gas System (TAGS). The TAGS project is comprised of three major components: a 36 inch, 800 mile long, buried chilled natural gas pipeline from Prudhoe Bay to Valdez, parallel to TAPS which will transport 2 BCF/D of natural gas; 10 gas compressor stations to facilitate transport of the gas through the pipeline; and an liquefied natural gas plant (to be designed for a capacity of 14 million tons of liquefied natural gas per year) and marine terminal to liquefy the gas and load onto liquefied natural gas tankers for shipment to Pacific Rim countries (Yukon Pacific Corporation[16], 1986). The estimated cost of this entire project are $8.6 billion (Treadwell[15], 1987). The operating costs for TAGS project are analogous to that of ANGTS (Economides and Ostermann[9], 1982).

Economic analysis (Economides and Ostermann[9], 1982) shows that the TAGS project is more attractive than the ANGTS project, yet still non-competitive compared to current world price of natural gas. On the other hand, a recent report on economic analysis (Anderson and Bukacek[17], 1987) based on Japan's Institute of Energy Economic (IEE) study shows that it is feasible to export 7 million tons/yr (50% capacity) to Pacific Rim countries by year 1998 and expand it to 14 million tons/yr (full capacity) of liquefied natural gas by year 2004. Currently, Japan imports 94% of its natural gas and mostly from Indonesia. The above study considers that by year 1998, Japanese demand alone will exceed 7 million tons/yr if crude oil prices reach $24/bbl and Korean demand for liquefied natural gas will increase by 3 million tons/yr beginning 1996 and will remain at annual level of 5 million tons until 2001. Taiwan may provide only a spot market for North Slope liquified natural gas. While, the Pacific Rim markets, supply/demand and economics clearly governs the fate of this project, other hurdles for this project are the current government policies and international political scene and financing. One big lift for this project is the recent approval of export of natural gas by President Reagan (Daily News Miner[18], Jan. 13, 1988). Other variation of this option is to transport liquefied natural gas via submarines to northeastern U.S. and Pacific Rim markets. This also faces problems of economic feasibility due to high transportation costs and the 1976 Alaska Natural Gas Transportation Act.

<u>Option 3: Conversion of Natural Gas to Methanol and/or Gasoline and Transporation through TAPS</u>

Marsden and co-workers (Marsden[19], 1979; Marsden[20], 1983, Corley and Marsden[21], 1987) proposed conversion of natural gas to methanol and transportation of crude oil-methanol mixture through TAPS. Because the mixture will be lower in viscosity than North Slope crude it will reduce the cost of pumping the mixture through TAPS. Methanol will be separated at Valdez and shipped to western U.S. markets. While this approach would eliminate the cost of new gas pipeline and has potential due to growing use of methanol as a vehicle fuel or octane enhancer in California & Colorado (OGJ[22], Dec. 14, 1987) there are large uncertainties associated with this approach. Firstly, environmental impacts due to use of methanol as fuel are not fully known. Secondly, the effect of mixing methanol with North Slope crude on the quality of ultimate crude needs to be investigated. Thirdly, the conversion of entire North Slope gas to methanol would triple the world output of methanol and cause significant depression in the methanol prices, unless the demand for methanol goes up in the western U.S. markets. Thus the economic picture is not clear. Conversion of methanol to gasoline is considered by many experts as another option, however, the conversion cost of this option is not favorable at this time. It may be also possible to combine the methanol option with other options i.e. only partially convert natural gas to methanol and use remaining North Slope gas for other options.

Option 4: Conversion of Natural Gas to Gasoline (One Step Process)

The option of conversion of natural gas to gasoline by one step process was first proposed in the recent Alaskan Natural Gas Utilization Workshop[23] (1987). Some experts believe that such a direct conversion process has potential for reduction in costs of manufacturing fuel for transport. While significant research is underway to develop technology to catalytically or noncatalytically convert natural gas to produce heavier hydrocarbons in the boiling range of gasoline/distillates. A significant time is necessary before this process can be commercially developed. Certainly detailed economic study and more research to develop the technology are needed.

Option 5: Natural Gas basedPetrochemical Industry

In 1979, Chem Systems[8], Inc. (1979) conducted preliminary study to investigate feasibility of a Petrochemical Industry in Alaska and potential of Pacific Rim, U.S. and Canadian markets for the products. This study recommended that such a petrochemical complex should concentrate on producing low and high density polyethylene, styrene, vinyl chloride and ethylene glycol. Apart from this study no other study has been done to investigate economics and market conditions. Lack of initiative and fear that such a petrochemical complex may not be feasible has caused delay in further studies and reduced the interest in this option.

Option 6: Natural Gas for Enhanced Oil Recovery

While, for the past decade, all the attempts to economically move the North Slope gas to markets have failed, the gas is currently being reinjected in the Prudhoe Bay Field. What did that gas reinjection do? The 1976 study of Prudhoe Bay Field (State of Alaska Report[24], 1976) shows that with sale of gas and crude oil production rate of 1.5 million B/D, the daily oil production rate would start to decline after recovery of 4.75 billion barrels in 1985 and without sale of gas it would start to decline after recovery of 5.75 billion barrels in 1987. So far 7.1 billion barrels of crude oil have been produced from Prudhoe Bay Field. The current Prudhoe Bay oil production rate is 1.5 million B/D. The recent predictions based on actual reservoir performance and history matching indicates that the possibility of producing 7.6 billion barrels of crude by end of the year 1991. Comparing the value of gas to oil, clearly the reinjected gas has served the best purpose.

At Prudhoe Bay, the bulk of the natural gas is compressed by powerful compressors and reinjected into the gas caps. This reinjection is beneficial in that it maintains high reservoir pressures, thus keeping the oil production rates relatively high. However, as the oil is produced, the gas cap expands, the reservoir pressure declines and GOR increases and the gas cycling occurs. This "gas cycling" has already been observed in some wells at Prudhoe Bay. The production rate of oil then may be limited by the compressor capacity. This however does not imply that the ultimate oil recovery will be adversely affected. However, the cost of reinjection will increase.

In the eastern and western portion of the Prudhoe Bay field the natural gas is also used for miscible flooding process. The heavier hydrocarbons are stripped from the natural gas and injected into the Prudhoe Bay Reservoir to achieve miscibility and thus enhance the oil recovery. Such an enhanced oil recovery process is certainly cost effective due to availability of natural gas. This is probably the lowest cost option for the North Slope gas utilization.

On the same token, the North Slope gas can be effectively utilized for the enhanced oil recovery in the other North Slope oil fields. Plans for undertaking miscible flood in the Kuparuk River Field (Kuparuk Formation) similar to Prudhoe Bay are already underway. Lisburne and Endicott fields also show promise for the use of natural gas for future enhanced oil recovery.

In addition, North Slope has huge resources of heavy oil and tar sands. In the Kuparuk River Unit the shallow West Sak Sands contain 15-25 billion barrels of heavy crude and the shallower Lower Ugnu Sands contain 6-11 billion barrels of tar sand bitumen and the Upper Ugnu Sands contain 5-8 billion barrels of tar sand bitumen. These resources are not recoverable by primary or secondary production methods. The use of natural gas as miscible flooding for enhanced oil recovery in West Sak and Ugnu reservoirs shows great promise. If thermal recovery methods are used for West Sak then the natural gas can be used as a boiler fuel for generation of steam or hot water. Steam can be complimented by addition of solvents such as heavier hydrocarbons which can be recovered from the natural gas. Hot WAG process consisting of injection of alternate hot water and solvents can also be designed. Prudhoe Bay gas has 12 percent CO_2 and can be recovered from natural gas. This CO_2 then can be used for miscible or immiscible gas drive.

Another EOR concept is based upon injection of mixture of steam and hydrogen for catalytic hydrogenation of heavy oil. This process can use naturally occuring trace elements as catalysts. The hydrogen required for the process can be derived from the North Slope natural gas.

While the type of EOR method for West Sak to be employed will depend upon the economics, certainly in most of these processes the use of North Slope gas shows the greatest promise among all the gas utilization options.

Currently, the esimates of amount of natural gas required for the miscible flood in the eastern and western portion of the Prudhoe Bay field which cover 2.2 billion barrel original oil in place are 200 MMSCF/D. The estimated incremental recovery of oil due to miscible flood is 115 MM barrels of crude, or 5.2% of the original oil in place (Alaska Oil and Gas Conservation Commission Report[25], 1984). Considering the life of this EOR project to be 10 years, the amount of natural gas required would be 0.73 TSCF. Considering extension of miscible flood to the entire field gives the estimate of natural gas required as 8 TCF. Applying same criteria for Kuparuk, Endicott and Lisburne fields for miscible flood the amount of natural gas requirements would be approximately 2, 0.5 and 1 TSCF, respectively.

Calculation of natural gas requirements for thermal recovery process such as steam injection for West Sak heavy oil show that approximately 18 TSCF of natural gas would be needed. These calculations assume the heating value of natural gas as 1000 Btu/SCF, efficiency of steam generator to be 70%, oil-steam ratio of 0.2, ultimate oil recovery of 30% of original oil in place and, energy content of steam to be 1160 Btu/lb.

These preliminary calculations suggest that there is sufficient natural gas on the North Slope to develop natural gas based EOR processes on the North Slope for Prudhoe Bay, Kuparuk, Lisburne, Endicott and West Sak reservoirs.

While the above options consider use of large amount of North Slope gas there are other options which use small amounts of North Slope gas and should also be considered. One such option is utilizing natural gas in the form of utilities for local usage. Gas turbine and steam boiler co-generator technolgies to convert natural gas into electricity are conventional and well proven commercially. The size of power plant facilities will be dictated by the local Alaskan market and economics. For large metropolitan areas like Fairbanks and Anchorage, utilization of North Slope gas by utilities does not look economical especially due to the fact that much closer Cook Inlet gas and Kenai gas fields can serve the same purpose for next 20-25 years. Other option that does not look economically attractive at the present is the use of natural gas in the form of fuel cells.

CONCLUSIONS

1. That the use of North Slope gas for the enhanced oil recovery on the North Slope fields is the best option.
2. It may be possible to convert part of the natural gas into methanol and transport it through the existing pipeline. The possibility of growing use of methanol as a vehicle fuel in California and Colorado makes this attractive option.
3. The option of transportation of natural gas in the form of liquefied natural gas to Pacific Rim markets such as Japan and Korea is currently not economical but may become potentially attractive in 1998 and beyond.

ACKNOWLEDGEMENT

The work on this paper was conducted and supported by the Petroleum Development Laboratory, University of Alaska, Fairbanks.

REFERENCES

1. Jones, H.P., and Speers, R.G.: *"Permo-Triassic Reservoirs of Prudhoe Bay Field, North Slope Alaska, in North American Oil and Gas Fields"*, AAPG Mem. Vol. 24, (1976) p. 23-50.

2. Young, A., Monaghan, P.H., and Schweisberger, R.T.: *"Calculation of Ages of Hydrocarbons in Oils-Physical Chemistry Applied to Petroleum Geochemistry I"*, AAPG Bull., Vol. 61, (1977) p. 573-600.

3. Seifert, W.K., Moldowan, J.M., and Jones, R.W.: *"Application of Biological Marker Chemistry to Petroleum Exploration"*, 10th World Petroleum Cong., (1979) p. 425-440.

4. Magoon, L.B., and Claypool, G.E.: *"Two Oil Types on North Slope of Alaska - Implications for Exploration"*, AAPG Bull., Vol. 65, (1981) p. 644-652.

5. Lewin and Associates: *"Handbook of Gas Hydrates, Properties and Occurrence"*, A report prepared for METC under Contract DE-AC21-82MC19239, November (1983).

6. Godbole, S.P., Kamath, V.A. and Economides, Ehlig-C.E.: *"Natural Gas Hydrates in the Alaskan Arctic"*, to be published in SPE Formation Evaluation, March (1988).

7. Roadifer, R.D., Godbole, S.P. and Kamath, V.A.: *"Estimation of Parameters for Drilling in Arctic and Offshore Environment in the Presence of Hydrates"*, SPE paper #16671, Proceedings of 62nd Annual Technical Conference and Exhibition of SPE, Dallas, TX, Sept. (1987).

8. Chem. Systems Inc.: *"Pacific Rim Market for Petrochemicals"*, A Report prepared for Royalty Oil & Gas Advisory Board, Department of Natural Resources, State of Alaska, October (1979).

9. Economides, M.J. and Ostermann, R.D.: "*Options for North Slope Gas Utilization*", A Report submitted to Division of Energy and Power Development, Lloyd Pernella, Director under Grant Contract #08-73-7-362, April (1982).

10. Corley, D.I. and Marsden, S.S. Jr.: "*An Evaluation of Options for Prudhoe Bay Natural Gas*", SPE Paper #12766, Proceedings of 1984 SPE California Regional Meeting held in Long Beach, CA, April 11-13 (1984).

11. Wadlington, C. Jr.: "*The Alaska Natural Gas Transportation System Perspective*", Proceedings of the Alaskan Gas Utilization Workshop, sponsored by the State of Alaska, U.S. Department of Energy and University of Alaska, Fairbanks, Alaska, June 30-July 2 (1987) p. 37-39.

12. Daily News Miner: "*What to Do with That Gas*", Nov. 18 (1987).

13. Jespersen, C.K.: "*Foothills Pipelines (Yukon) Ltd. Perspective*", Proceedings of the Alaskan Gas Utilization Workshop, sponsored by the State of Alaska, U.S. Department of Energy and University of Alaska, Fairbanks, Alaska, June 30-July 2 (1987) p. 50-52.

14. Daily News Miner: "*ARCO Withdraws From Gas Pipeline Project*", Dec. 14 (1987).

15. Treadwell, M.: "*Summary - The Trans Alaska Gas System Perspective*", Proceedings of the Alaskan Gas Utilization Workshop, sponsored by the State of Alaska, U.S. Department of Energy and University of Alaska, Fairbanks, Alaska, June 30-July 2, (1987) p. 40.

16. Yukon Pacific Corporation: "*Trans Alaska Gas System - Project Description*", A Proposal prepared by YPC, December (1986).

17. Anderson, P.J. and Bukacek, R.F.: "*Evaluation of The Feasibility of ExportingNorth Slope Alaska Gas as LNG*", Inst. of Gas Tech. Project Report 40268, June (1987).

18. Daily News Miner: "Approval of Export of Natural Gas", Jan. 13, (1988)

19. Marsden, S.S. Jr.: "*A Proposal for Transportation of Prudhoe Bay Gas to Market*", SPE Paper #8296, Proceedings of SPE Annual Meeting held in Las Vegas, Nevada, September (1979).

20. Marsden, S.S. Jr.: "*Methanol - A Viable Energy Source in Today's World*", Annual Review of Energy, Vol. 8 (1983) p. 333-354.

21. Marsden, S.S.: "*Transportation of Prudhoe Bay Natural Gas and Heavier Crude Oils to Market*", Proceedings of the Alaskan Gas Utilization Workshop, sponsored the the State of Alaska, U. S. Department of Energy and University of Alaska, Fairbanks, Alaska, June 30-July 2, (1987) p. 114-118.

22. Oil and Gas Journal: "*Methanol Fuel Programs Move Ahead in 2 States*", Dec. 14 (1987) p. 20.

23. "Proceedings of the Alaska Natural Gas Utilization Workshop" sponsored by the State of Alaska, U.S. Department of Energy and University of Alaska, Fairbanks, Alaska, June 30-July 2, (1987)

24. Alaska Oil and Gas Conservation Commission Report: "*Prediction of Reservoir Fluid Recovery Sadlerochit Formation, Prudhoe Bay Field*", January (1976).

25. Alaska Oil and Gas Conservation Commission Report: *"Prudhoe Bay Miscible Gas Project"*, Nov. (1984).

26. Private Communications: Alaska Oil and Gas Conservation Commission, December (1987).

COALESCENCE BEHAVIOR OF WATER-IN-OIL EMULSIONS

E. E. Isaacs, H. Huang, R.S. Chow and A.J. Babchin

Alberta Research Council
Oil Sands and Hydrocarbon Research Department
P.O. Box 8330, Postal Station F
Edmonton, Alberta T6H 5X2 Canada

In this paper the use of electroacoustic techniques involving the application of a sonic field and the detection of an electric field, for monitoring coalescence of water droplets in non-polar media will be discussed. This technique was used to evaluate the rate and extent of dewatering in oil continuous emulsions when surface active chemicals were added. The results showed that a combination of an oil soluble demulsifier and water soluble surfactant was substantially more effective in causing droplet coalesence than the individual components. An explanation for these findings were based on studies of time-dependent interfacial tensions at the oil/water interface and electrokinetic properties. The results indicated that a direct relationship exists between the adsorption behavior at the oil/water interface (apparent rate of spreading) and emulsion stability.

INTRODUCTION

General Remarks

The stability of water-in-crude oil emulsions and the factors contributing to that stability is a long standing problem of industrial importance in the production of oil from underground reservoirs. Although there has been a great deal of effort expended in the investigation of the destabilization of oil continuous emulsions, the actual mechanisms are still not well understood[1]. Indigenous material in the oil tend to gather at the interface and play a significant role in hindering the thinning and rupture of the liquid films and act as a structural barrier to coalescence[2,3,4]. The complex chemical nature of crude oils makes it difficult to relate the dispersion behavior to the physico-chemical properties at the crude oil-water interface. In addition, the non-polar and non-transparent nature of the oleic phase is a significant obstacle for studies of the interactions of the suspended water droplets in real systems.

Electroacoustic Measurements Applied to Non-Polar Medium

The recent development of electroacoustical techniques has shown considerable promise for electrokinetic measurements of colloidal systems and the direct monitoring of the rate and extent of coagulation (flocculation and coalescence) of water droplets in non-transparent oil continuous media[5,6]. The electroacoustic measurement for colloidal systems in non-polar media is based on the ultrasound vibration potential (UVP) mode which involves the application of a sonic field and the detection of an electric field. A schematic diagram of the probe and the principle of UVP are represented in Figure 1. When a voltage U2 is applied at the transducer a sound wave propagates into the colloid. If the density of the disperse and the continuous phases differ, relative motion between the colloidal particles and their double layer will result. The combined relative motion will generate an electric field, which is detected as a voltage U1, between the electrodes. The measured signals are proportional to the high frequency electrophoretic mobility $\mu(\omega)$. As derived by Babchin et al.[5], the frequency electrophoretic mobility, for the case of low potentials, can be expressed by:

$$\mu(\omega) = \frac{|U_o|}{E_o} = \frac{\epsilon \varsigma f(\kappa R)}{\sqrt{(6\pi\eta\bar{R})^2 + \left[\frac{4}{3}\omega \rho_{eff} R^2\right]^2}} \qquad [1]$$

with

$$\bar{R} = \frac{R_{eff}}{R} = 1 + \frac{R}{\delta} \qquad \rho_{eff} = \rho_o + \frac{9}{4R}\sqrt{\frac{2\eta\rho}{\omega}}\left(1 + \frac{2R}{9\delta}\right) \qquad [2]$$

where ε, η, and ρ are the dielectric permittivity, viscosity, and density of the continuous phase respectively. ρ_o is the density of the particle, ς is the electrokinetic potential, $f(\kappa R)$ is the Henry function, κ is the Debye-Huckel function that characterizes the extension of the double layer. R is the particle radius and ω is the frequency.

The supplementary phase angle $\phi(\omega)$ between the applied electric field and the particle velocity response, at a fixed frequency ω, is given by:

$$\tan\phi = \frac{\mathrm{Im}\, U_o}{\mathrm{Re}\, U_o} = -\frac{2}{9}\frac{\omega R^2 \rho_{eff}}{\eta\bar{R}} \qquad [3]$$

The magnitude of the potential difference between the electrodes, $\Delta\Psi_o$ in the circuit U1 is given by:

$$\mathrm{UVP}(\omega) = \frac{\Delta\Psi_o}{V_o} = \frac{\phi\Delta\rho c G_f}{K^*}\mu(\omega) \qquad [4]$$

Equations 1-4 clearly show that an increase in the effective particle radius, promoted by a coagulation process, will result in the diminution of the UVP signal and a shift in the phase angle. In addition, the low value of the complex conductivity of oil, K^* acts as a natural amplifier to provide for a significant $\Delta\Psi$ that makes it easy to monitor UVP even for small values of $\mu(\omega)$. By monitoring both the UVP signal and phase angle ϕ it is possible to distinguish between changes in ς which effect the UVP signal only and changes in particle radius which effect both the UVP and $\phi(\omega)$[7].

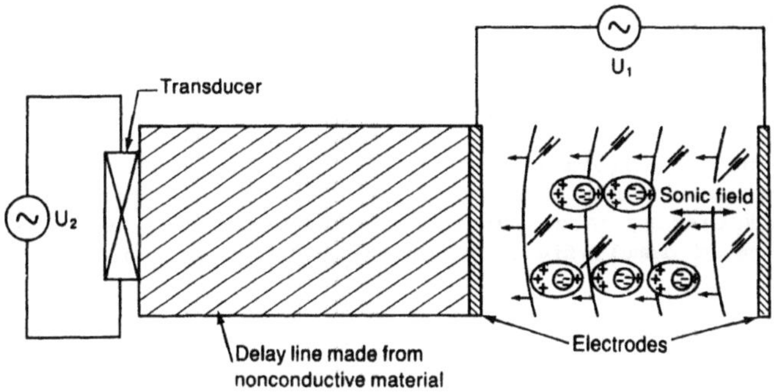

Figure 1 **Schematic diagram showing the principles of Ultrasound Vibration Potential (UVP).**

Adsorption at the Oil/Water Interface

In order to enhance the coagulation process it is common to use chemical demulsifiers which are believed to:

a) promote the flocculation of the droplets by weakening the repulsive forces that stabilize the emulsion; and

b) enhance the drainage of the interfacial film between the flocculated droplets.

The choice of chemical is usually based on trial and error procedures making demulsifier technology more of an art than a science. In most cases a combination of chemicals are used in the demulsifier formulation so as to achieve both efficient flocculation and coalescence. The type of demulsifiers and their effect on interfacial area are among the important factors which influence the coalescence process. Time-dependent interfacial tensions have been shown to be sensitive to these factors[8-10]. The relation between time-dependent interfacial tensions and the adsorption of surfactants at the oil/aqueous interface was considered by a number of researchers[11-19]. From studies of the time-dependent tensions at the interface between organic solvents and aqueous solutions of different surfactants, Joos and co-workers[13-17] concluded that the adsorption process of the surfactants at liquid/liquid interface was not only diffusion controlled but that adsorption barriers and the reorientation of surfactant molecules were important mechanism depending on the system. For the case of surfactant adsorption from the oil phase to the oil/water interface, it was thought that a reorientation process at the interface was the rate controlling step[17]; adsorption occurred at a much slower rate than that observed for a purely diffusion controlled situation. Vogler[11] developed a mathematical model to derive semiquantitative kinetic parameters interpreted in terms of transport and adsorption of surfactants at the interface. The model was fitted to experimental time-dependent interfacial tension and empirical models of concentration-dependent interfacial tension were compared to theoretical expressions for time-dependent surfactant concentration. Adamczyk[12], theoretically related the mechanical properties of the interface to the adsorption kinetics of surfactants by introducing the compositional surface elasticity which was defined as the proportionality coefficient between arbitrary surface deformations and the resulting surface concentrations. Although the expressions used to describe the adsorption process differed from one another, it was demonstrated that the time-dependent interfacial tensions mirrored the change of surface-active substances at the interface.

For studies with real systems, we have used the simplified approach of examining changes at the oil/water interface without specifying adsorption mechanisms or pathways. Based on measurements of time-dependent interfacial tensions, the following expression (termed the apparent spreading rate) served to characterize the relative adsorption performance of demulsifiers or demulsifier combination:

$$\text{Spreading rate} = \gamma_o - \gamma_{s/e} / \Delta t \qquad [5]$$

where γ_o and $\gamma_{s/e}$ are the steady-state value of oil/aqueous interfacial tension in the absence and presence of added chemical respectively; Δt represents the time required to reach the steady-state tension, $\gamma_{s/e}$. The determination of this parameter from experimental data is shown in Figure 2.

In this paper we examine the role of mixed surfactants in the demulsification of water-in-Leduc oil emulsion by application of the spreading rate method which is then correlated with the electroacoustic results and centrifugation. Microelectrophoresis using the reverse emulsion was also used to investigate the adsorption process. The results show both a very good correspondence between the various techniques and provide insight on the synergistic adsorption behavior of the hydrophobic and hydrophilic surfactants.

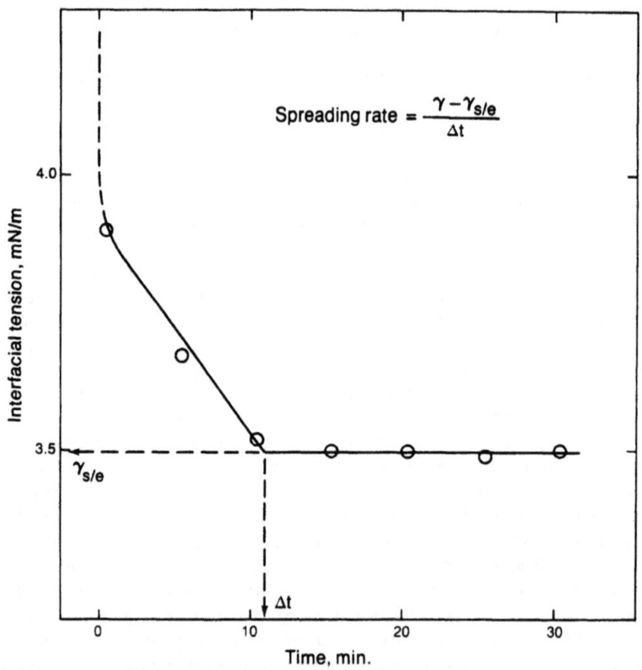

Figure 2 **Evolution of interfacial tension data showing the method for determining the spreading rate parameter.**

EXPERIMENTAL PROCEDURES

Materials

The aqueous phase used was either distilled water or $1x10^{-2}$ M NaCl solution with pH adjusted with NaOH or HCl. Crude oil from the Leduc field was obtained from the Alberta Research Council's sample bank. The physical properties of this oil had been reported previously[20]. The chemicals used in this study were:

a) Aerosol OT (dioctyll ester of sodium sulfosuccinic acid) having a molecular weight of 444 was supplied by Fisher Chemicals as a 75% active concentrate in aqueous solution; and
b) Duomeen C (n-alkly-1,3-propanediamine) supplied by AKZO Chemicals. The hydrocarbon chain length ranged from C_6 to C_{18} and consisted mostly of C_{12} and C_{14}.

Both chemicals were used as received.

Preparation of Water-In-Oil Emulsions

The volume of the emulsion was kept constant at 100 cm^3. Measured volumes of crude oil and the aqueous phase were initially mixed by stirring for 5 min using a magnetic stirrer. This was followed by five min of sonication (Fisher model 300 Sonic Dismembrator) using a medium probe and the power set at 50% of the maximum output. The choice of sonication time was made on the basis of previous work[6]. The emulsions thus formed were stable for a period of several days.

Monitoring of Emulsion Coalesence

The Electrokinetic Sonic Analysis System (Matec Instruments, Boston) was used to monitor the rate of demulsification. The prepared water-in-oil emulsions were circulated through the parallel plate transducer (Model PPL-80; Matec Instruments) and the UVP signal measured as a function of time. As shown in Figure 3, the UVP signal increases with increasing water content of the emulsion. No signal is obtained for Leduc oil in the absence of emulsified water. The non-linearity of the plot may reflect the difficulty in making a consistent droplet-size distribution for different initial water content. The UVP signal in the absence of added chemicals remained essentially unchanged as a function of time. The kinetics of coagulation (flocculation and/or coalescence) caused by addition of demulsifier was monitored by the change in the UVP signal. The change in UVP signal was also confirmed by a change in phase angle (see equation 3) in order to differentiate changes in UVP due to coalescence processes from those due to purely changes in electrokinetic potential (see equation 1).

After the electroacoustic measurements were finished, the sample was placed into a centrifuge tube and spun at 1500 rpm in a table top centrifuge (International Model HN). The amount of released water was recorded to confirm the relative dewatering efficiency of the demulsifiers.

Time-Dependent Interfacial Tensions

The interfacial tension measurements were carried out using the spinning drop tensiometer using methodology described in previous work[21,22]. The main advantage of using this technique was the ease and convenience of operation and its amenability for further studies planned at elevated temperatures. The major limitation of this technique is that adsorption phenomena below about 60 sec are not captured because of the

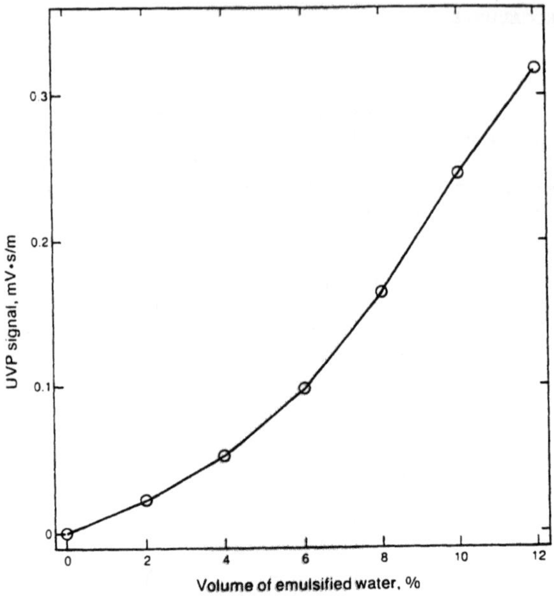

Figure 3 Sensitivity of UVP signal to water content in the emulsions.

time required to introduce the droplet into the tube and acquire the initial tension. Our experience suggests that the method can be successfully used with crude oil systems where adsorption processes are not diffusion controlled; the presence of surfactants indigenous to the oil creates barriers to adsorption.

In all experiments the Duomeen C was added in the desired concentration to the Leduc crude, while Aerosol OT was added to the aqueous phase.

The oil drop volume used was 5 μl, introduced into a 4 mm i.d. glass capillary containing the aqueous solution. The tube was then sealed with a tightly fitting silicon-rubber septum. A teflon screw was used to keep the septum tight after the capillary tube was inserted into the shaft of the tensiometer. The capillary tube was accelerated to a predetermined angular velocity chosen from preliminary measurements. The angular velocity was selected to produce drops with the length about 4 times greater than the diameter, this simplifies the calculation of interfacial tension. With experience the length of time to obtain the initial interfacial tension measurement corresponded to about 60 sec. The value of interfacial tension was recorded periodically until the steady-state tensions were reached. A schematic of the technique is shown in Figure 4 together with the depiction of the adsorption of Duomeen C from the oil to the oil/water interface and Aerosol OT from the water to the water/oil interface. The molecular structure of the two additives are also given in Figure 4.

Preparation of Crude Oil-In-Water Emulsion

For electrophoresis measurements, a 5 g sample of Leduc oil was mixed in 100 ml of water containing 0.01 M NaCl. The pH of the water was adjusted using NaOH (pH of 9 to 11) or HCl (pH = 5 to 3). The mixture was passed through a hand-held homogenizer (Chase Logman Corp., Hixville,

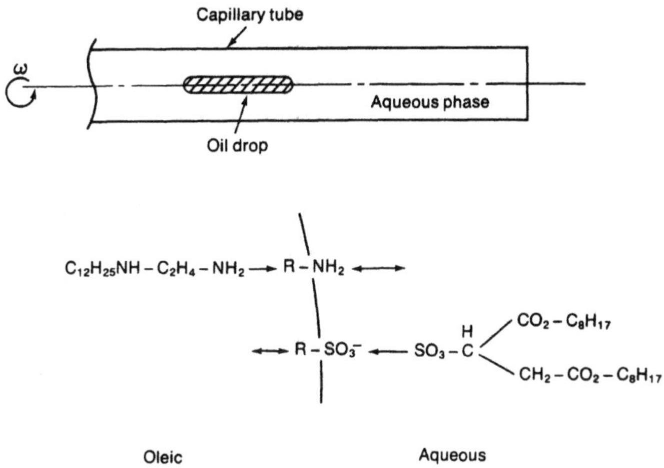

Figure 4 Schematic diagram of the spinning drop capillary depicting the adsorption of Duomeen C (oil soluble) and Aerosol OT (water soluble) from the bulk to the interface.

NY) three times. Nearly all of the crude oil was dispersed in the water, giving particle radii ranging between 1 and 8 μm. For oil droplets containing Duomeen C, 100 ppm of the demulsifier was added to the oil prior to emulsification.

Electrophoretic Mobility Measurements

The measurements were conducted using the Rank Brothers Particle Microelectrophoresis Apparatus Mark II equipped with a flat cell. Samples were prepared by diluting a 100μl sample of the oil-in-water emulsion with 20 ml of 10^{-2} M NaCl containing Aerosol OT (0 to 100 ppm). In experiments with Duomeen C, the reagent was added to the oil phase in the desired amount prior to emulsification. The pH was adjusted using HCl or NaOH to cover the range 2.5 to 11.5. All experiments were conducted at 30°C.

RESULTS AND DISCUSSION

Effect of Demulsifier Mixture

As shown in a previous study[6], Duomeen C which was effective in causing flocculation of the water droplets, was not very effective in breaking the interfacial film formed between the water droplets, which inhibits coalescence. However, Duomeen C in combination with Aerosol OT, a hydrophilic surfactant, was much more effective in causing water separation compared to the individual chemicals. This effect is shown in Figure 5 for a 6 vol.% water-in-Leduc crude emulsion where both the UVP signal (20 min after chemical addition) and the volume of water recovered by centrifugation are plotted against the weight percent of Duomeen C in the mixture. As expected Aerosol OT, the water-soluble surfactant by itself had practically no effect on either the UVP signal or the water separation. Duomeen C alone also had little effect on the amount of water recovered by centrifugation. The change in UVP signal therefore, likely reflected Duomeem C's ability to flocculate the droplets. The mixture of the two chemicals, however, performed in a synergistic manner

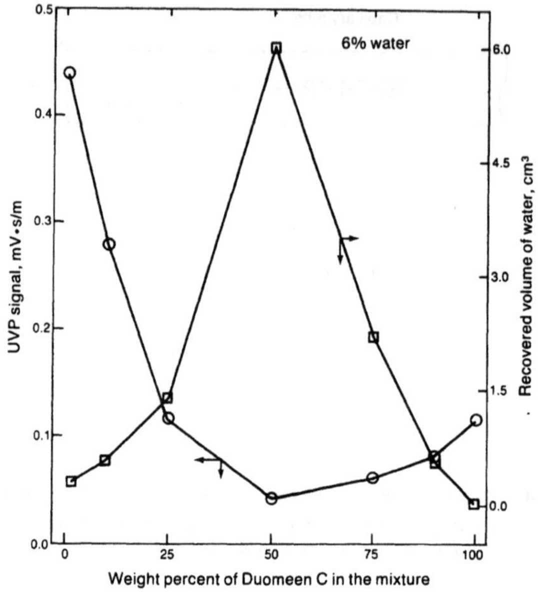

Figure 5 **Comparison of the coalesence process using mixtures of Duomeen C and Aerosol OT (total conc. is 100 ppm) as determined by electroacoustic analysis (o) and the centrifuge (□). The UVP signal is taken at 20 minutes after addition of demulsifiers.**

with a 1:1 mixture of chemicals being most effective. It is also apparent that there is a direct correspondence between the minimum in UVP signal and maximum in water recovery by centrifugation. Recent work by Sjobolom et al[9,23] has also found that a combination of chemicals such as amine and medium chain alcohol, speeded up the destabilization of water-in-Norwegian crude oil emulsion. The alcohol was thought to modify the rigidity of the film, while the amine was shown to interact with the interfacial groups.

Dynamic Interfacial Tensions During Adsorption at Interfaces

In order to understand the above phenomena and, especially, the reasons for the dewatering effectiveness resulting from the interactions between the two surfactants, time-dependent interfacial tensions were measured to examine the transfer of the surfactants from the bulk to the interface. Figure 6 shows the evolution of oil/water interfacial tension for the case of no additive, Duomeen C (100 ppm) added to the oil phase, and Aerosol OT (100 ppm) added to the aqueous phase. The slight increase in crude oil/aqueous interfacial tension with time when no additive is added, is indicative of the transfer of indigenous surfactant present in the oil to the aqueous phase which as has been shown by Rubin and Radke[18], is accentuated by the large volume ratio of the aqueous to oleic phase inherent in the spinning drop method. Duomeen C transferring from the oil phase to the oil/water interface required about 30 min to reach steady-state. This long time-effects seem to be quite general for surfactants solved in the oil phase and nearly insoluble in the aqueous phase[17]. Van Hunsel et al.[15,17], have argued that the long time effects cannot be explained by simple diffusion[24,26] but likely molecular reorientation processes at the interface control the adsorption kinetics.

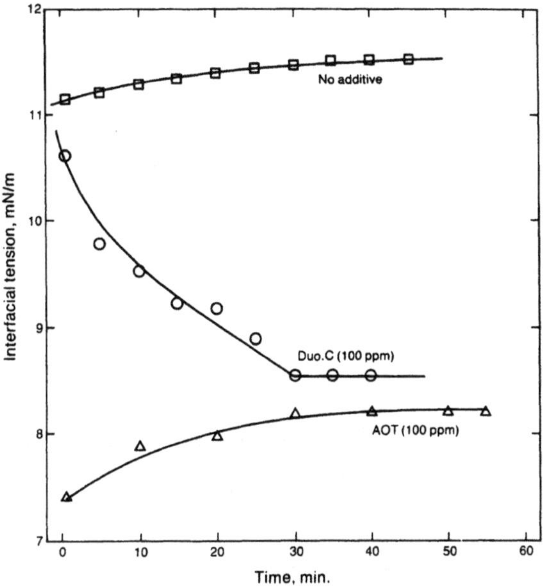

Figure 6 **Evolution of interfacial tension for Leduc crude oil (□), 100 ppm Duomeen C in the oleic phase (○), and 100 ppm Aerosol OT in the aqueous phase (△).**

In the present work the adsorption mechanism is even more complicated because in addition to molecular orientation the crude oil/water interface is already saturated by naturally present surface active material which may have to be displaced from the interface before adsorption can take place.

In marked contrast, the transfer of Aerosol OT from the aqueous to the aqueous/oil interface appears to be instantaneous and cannot be completely captured in the time scale of the spinning drop measurements. This fast adsorption time-scale is anticipated for surfactants solved in the aqueous phase and partition to a significant extent into the oil phase below the critical aggregation number[27]. A small amount of desorption of material from the interface is suggested by the slight increase in tension with time (Figure 6).

For the case of mixtures of Duomeen C and Aerosol OT, Figure 7 shows that the presence of even 10 ppm Aerosol OT in the mixture considerably reduces the time to achieve steady-state tension, $\gamma s/e$. As with coagulation studies, a 1:1 mixture of the reagents proved to be most effective in the rate and extent of tension reduction (data measured for other mixtures have not been included in Figure 7 for clarity). It is also apparent that the mixture is considerably more surface active than the individual components. It is not known, however, if association between the amido and sulfonate groups of the two surfactant molecules is taking place at the interface; this is a subject of further investigation from this laboratory.

Apparent Spreading Rate

Based on measurement of time-dependent interfacial tension, we have defined an apparent spreading rate as given in [5], which served to

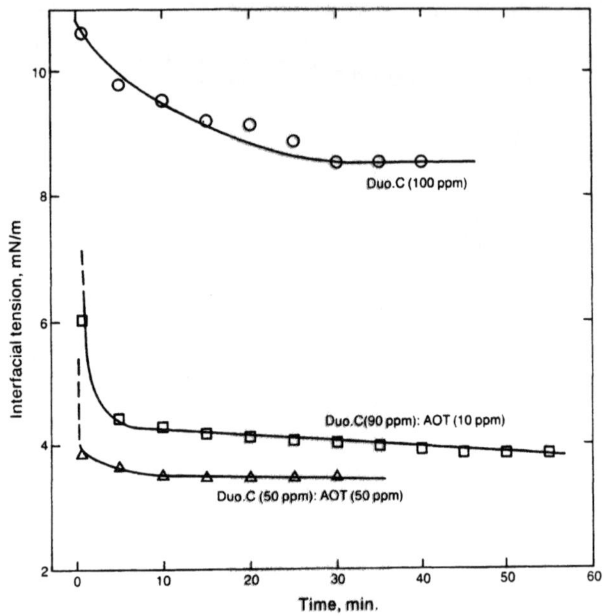

Figure 7 Evolution of interfacial tensions for Leduc oil with varying ratio of Duomeen C / Aerosol OT (concentration in ppm): 100/0 (○), 90/10 (□), and 50/50 (▲).

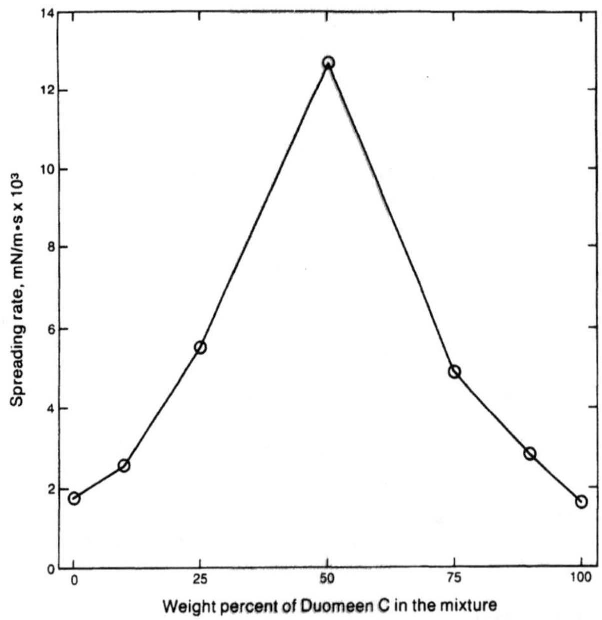

Figure 8 Apparent spreading rate as a function of the ratio of Duomeen C and Aerosol OT concentrations in the mixture. The total concentration is 100 ppm.

characterize the adsorption process. The spreading rate is a measure of both the ease of deformation of the interface and the speed of adsorption or mass transfer of material to the interface. Figure 8 shows a plot of the spreading rate as a function of the ratio of Duomeen C and Aerosol OT concentrations in the mixture. It is readily evident that the maximum spreading rate occurs at 1:1 ratio of the reagents. Figure 9 shows that there is an excellent agreement between the spreading rate and both the water recovery by centrifugation and the final UVP signal. The direct correlation between results of dynamic interfacial tensions and the results of coalescence or dewatering efficiency has not to our knowledge, been reported.

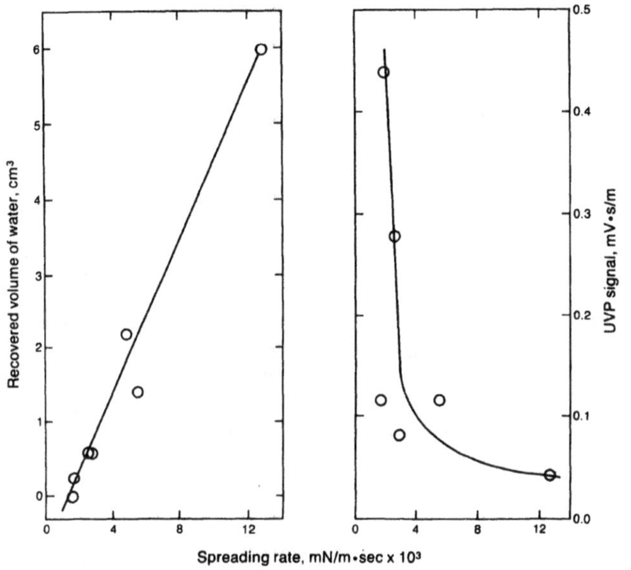

Figure 9 **Comparison of electroacoustic analysis and dewatering efficiency of the mixed demulsifier as a function of the apparent spreading rate.**

Electrokinetic Properties

In conjunction with the above studies of the time-dependent interfacial tension behavior and the coalescence properties of water-in-Leduc crude oil emulsions, the electrokinetic behavior of oil droplets in aqueous media (reverse emulsion) was also studied in the presence of Duomeen C and Aerosol OT. It was hoped that this will provide additional information on the adsorption of material at the oil/water interface.

Figure 10 gives a plot of electrophoretic mobility versus pH in the bulk phase in the presence of 10^{-2} M NaCl. In the absence of any additives (curve (a)), the mobility of the droplets varied from positive at acid pH to negative at basic pH with an isoelectric point of about pH = 7. The surface charge of the crude oil interface is derived from the pH dependent dissociation of functional groups naturally present in the crude such as: carboxylic (pKa~4.5) and sulfonic moieties (pKa~2) which are active at basic pH and are responsible for the negative charge; and amine groups (pKa~9) which are active at acidic pH and are responsible for the positive charge at the surface[28].

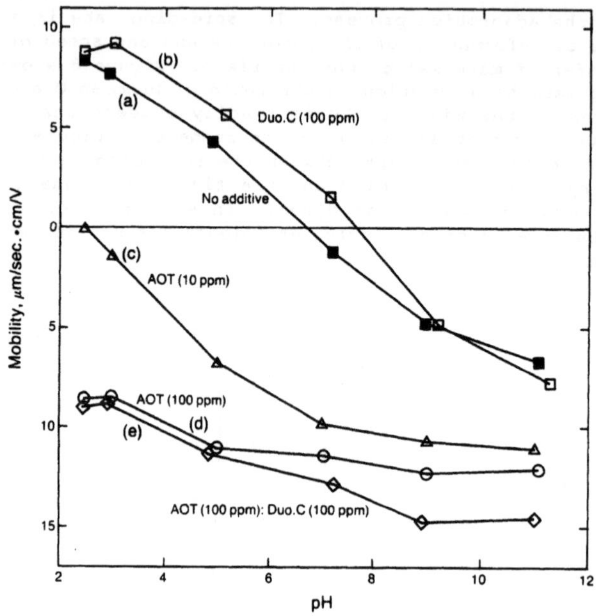

Figure 10 **Electrophoretic mobility of crude oil droplets as a function of bulk pH in aqueous systems: (a) no additive; (b) 100 ppm Duomeen C in the oil; (c) 10 ppm Aerosol OT; (d) 100 ppm Aerosol OT; and (e) mixture of 100 ppm Duomeen C in the oil and 100 ppm Aerosol OT in the aqueous phase.**

The addition of Duomeen C to the oil phase prior to making the emulsion, has the effect (curve (b)) of shifting the mobilities to more electropositive values with the isoelectric point now at pH ≅ 8. This is an indication that Duomeen C is adsorbing at the interface and changing the balance of anionic and cationic moieties. When even 10 ppm of the anionic Aerosol OT was added to the aqueous phase, the mobility of the droplets became negative at all values of pH (curve (c)). Addition of 100 ppm Aerosol OT, increased the shift towards negative mobilities in the acidic region to the extent that the mobilities were practically independent of pH (curve (d)). This indicated that Aerosol OT has strong capability to adsorb at the interface and either displace the indigenous amine molecules from the interface or form a complex with these groups thereby neutralizing their influence. When both surfactants were used during the measurement (curve (e)), the behavior was similar to that of Aerosol OT by itself (curve (d)). This is perhaps another indication of the ability of Aerosol OT to 'neutralize' the influence of amine-type moieties at the oil/water interface.

While the electric properties behavior shown in Figure 10 is an interesting study in it own right, it is not possible to directly correlate the results with those for coalescence and apparent spreading rate, especially since we are considering the reverse emulsions. However, as will be discussed below, the effect of pH on dissociation of groups at the interface should provide indicators for the conditions at which adsorption of the demulsifiers would be favored.

Effect of pH on Spreading Rate

The electric properties study provided good indication that pH of the aqueous phase can influence the dissociation and adsorption of both natural and added surfactants at the interface. This in turn is expected to influence the performance of the demulsifiers and the stability of water-in-oil emulsion. Figure 11 provides a comparison of the change in the apparent spreading rate values, from dynamic interfacial tension measurements, and the results of dewatering efficiency for 10 vol.% water-in-Leduc crude emulsions as a function of the initial pH (constant ionic strength of 10^{-2} M NaCl) of the dispersed aqueous phase. An Aerosol OT to Duomeen C ratio of 1:1 with a total concentration of 100 ppm was used in all experiments. As before, the very good correspondence between the coalesence efficiency and the apparent spreading rate is clearly evident. At high (above 9) and low pH (below 4), the low spreading rate values accompanied by poor water recovery from centrifugation represented the emulsions being stable. In the 5<pH>9 range, both the apparent spreading rate and the dewatering efficiency reached peak values. This suggest that at about neutral pH conditions the stabilizing influence of the polar groups of the natural surfactants can be eliminated as evident from the electric properties (curve (a) of Figure 10). Thus, it appears that the neutral pH conditions represent the most favorable conditions for the transfer to the interface of the Duomeen C-Aerosol OT mixture (or the complex formed by reaction of these reagents) and the optimum performance results.

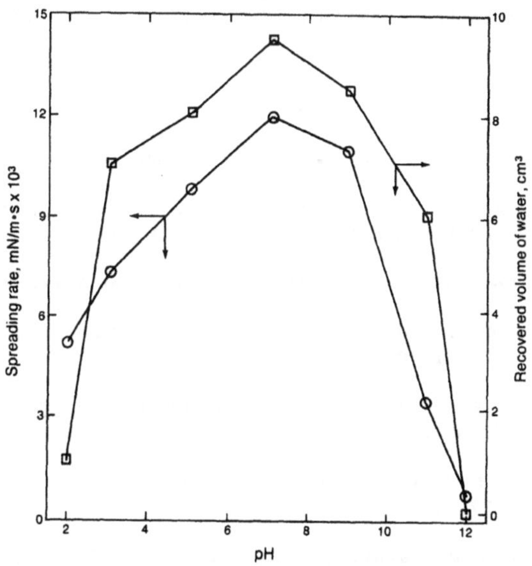

Figure 11 **Comparison of the effect of bulk pH on apparent spreading rate and water recovery by centrifugation for a 10 vol.% water-in-Leduc crude emulsions. A ratio of Duomeen C to Aerosol OT of 50 ppm:50 ppm was used in all experiments.**

CONCLUSIONS

An electroacoustic method based on the application of a sonic field and detection of the electric signal proved to be a reliable technique for measuring the kinetics of coagulation/coalesence process of colloidal systems in non-polar and non-transparent media like water-in-crude oil emulsion.

Interfacial tensiometry is technique which is sensitive to the adsorption of surface-active solutes. The time-dependent interfacial tension measurements served to reveal the physico-chemical changes at oil/aqueous interface during mass transfer of demulsifiers and demulsifier mixtures to the interface. The apparent spreading rate which was determined from the time-dependent tensions provided a measure for the ease of deformation of the interface and the speed of adsorption to the interface. The apparent spreading rate parameter correlated very well with the coalescence behavior and dewatering efficiency of the emulsion.

Electrokinetic properties of crude oil droplet dispersed in the aqueous phase provided useful information about the charge groups of the indigenous surfactants in the crude oil which likely act as barriers to the adsorption of added demulsifiers. The pH (and likely ionic content) of aqueous dispersed phase, therefore, plays an important role in the process of droplet coalescence in these systems.

ACKNOWLEDGEMENTS

We thank the Alberta Research Council and the Academia Sinica for financial support. We thank Kelly Cymbala for help in typing this manuscript. Alberta Research Council Contribution No. 2024.

REFERENCES

1. E.J. Johansen, I.M. Skjarvo, T. Lund, J. Sjoblom, H. Soderlund and G. Bostrom, Colloids and Surfaces, 34:353 (1988/1989).

2. Th.F. Tadros and B. Vincent, Encyclopedia of Emulsion Technology, Ch. 3, P.Becker (ed.), Marcel Dekker, New York, (1984).

3. K. Lissant, Demulsification, Marcel Dekker, New York, (1983).

4. T.J. Jones, E.L. Neustadter and K.P. Wittingham, J. Can. Pet. Technol., 17:100 (1978).

5. A.J. Babchin, R.S. Chow and R.P. Sawatzky, Adv. Colloid Interface Sci., 30:111 (1989).

6. E.E. Isaacs, H. Huang, A.J. Babchin and R.S. Chow, Colloids and Surfaces, 46:177 (1990).

7. A.J. Babchin, R.P. Sawatzky, R.S. Chow, E.E. Isaacs, and H. Huang, presented at the 21st Annual Fine Particle Society Meeting, San Diego, August, (1990).

8. K.C. Taylor and L.L. Schramm, Colloids and Surfaces, 47:245 (1990).

9. J. Sjoblom, Li Ming-yuan, H. Hoiland and E.J. Johansen, Colloids and Surfaces, 46:127 (1990).

10. C.I. Chiwetelu, V. Hornof and G.H. Neale, AIChE Journal, 36(2):233 (1990).

11. E.A. Vogler, J. Colloid Interface Sci., 133:228 (1989).

12. Z. Adamczyk, J. Colloid Interface Sci., 133:23 (1989).

13. M. Vermeulen and P. Joos, Colloids and Surfaces, 36:13 (1989).

14. M. Vermeulen and P. Joos, Colloids and Surfaces, 33:337 (1988).

15. J.V. Hunsel and P. Joos, Colloids and Surfaces, 25:251 (1987).

16. J.V. Hunsel and P. Joos, Colloids and Surfaces, 24:139 (1987).

17. J.V. Hunsel, G. Bleys and P. Joos, J. Colloid Interface Sci., 114:432 (1986).

18. E. Rubin and C.J. Radke, Chemical Engineering Science, 36:1129 (1980).

19. J.M.G. Lankveld, and J. Lyklema, J. Colloid Interface Sci., 41:454 (1972).

20. J.S. Buckley, K. Takamura, and N.R. Morrow, SPE Reservoir Eng., 4:332 (1989).

21. E.E. Isaacs and K.F. Smolek, Can. J. Chem. Eng., 61:233 (1983).

22. E.E. Isaacs, J.D. Maunder, and Li Jian, ACS Symposium Series, Borchardt and Yen, ed., 9 (1989).

23. J. Sjoblom, H. Soderlund, S. Lindblad, E.J. Johansen, and I.M. Skjarvo, Colloid & Polymer Sci., 268:389 (1990).

24. A.F.H. Ward and L. Tordai, J. Chemical Phys., 14:453 (1946).

25. A.F.H. Ward and L. Tordai, Nature, 154:146 (1944).

26. K.S.G. Doss, Kolloid Zeits, 84:138 (1938).

27. R. Aveyard, B.P. Binks, P.D.I. Fletcher, and J.R. Lu, J. Colloid Interface Sci., 139:128 (1990).

28. K. Takamura and R.S. Chow, Colloids and Surfaces, 15:35 (1985).

EFFECT OF FLOCCULATION ON GYPSUM FILTRATION EFFICIENCY

B.M. Moudgil and Shou-Lian Zhu

Mineral Resources Research Center
Department of Materials Science & Engineering
University of Florida
Gainesville, FL 32611

ABSTRACT

Improvement in filtration efficiency of gypsum upon flocculation is determined to be due to reduction in the blockage of filter media pores and in the filter cake resistance. An optimum in polymer dosage is explained in terms of stabilization of fines and deformability of the flocs formed. Anionic polyacrylamides were found to be more effective filter aids than cationic polymers.

INTRODUCTION

About one and a half tons of gypsum is produced for each ton of phosphate rock digested to produce phosphoric acid. Separation of gypsum from phosphoric acid is achieved by vacuum filtration using pan filters. Fluctuations in filtration efficiency are not uncommon and depend on the amount of impurity elements in the feed rock and operation conditions. However, the exact role of the impurities is not yet completely understood. In any case, higher filtration rates and low soluble losses of P_2O_5 in the filter cake are highly desired. Improvement in filtration efficiency by flocculation of gypsum fines has led to the development of "filter aids". It has been reported that the effect of filter aids on filtration efficiency has not always been positive. Inconsistencies in the performance of filter aids are generally attributed to lack of fundamental knowledge about the role of polymers in the flocculation process. In this study, phosphate rock from the Florida Phosphate Mining District was used to evaluate polymers as filter aids. Attempts were also made to understand the underlying mechanisms.

BACKGROUND

In developing the filtration rate equation, it is assumed that the velocity of fluid through cake pores obeys the Hagen-Poiseuille Law[1].

$$Q = \frac{\pi C}{8\eta} \frac{\Delta P}{H} \gamma^{4M} \qquad (1)$$

where:

Q = Flow rate (m^3/Sec); C = Constant; P = Pressure (kPa); η = Viscosity (kPa•sec);
H = Bed depth (m); γ = Radius of capillary (m); M = Compressibility coefficient

Based on the above equation, several investigators notably Ruth, Sperry, Lewis, Carman and Bonilla, developed the filtration equations under the conditions of both constant pressure and constant filtration rate[2,3,4]. The general form of the equation developed was as follows.

$$Q = \frac{A \cdot \Delta P}{\eta\left(\frac{\alpha W V}{A} + R\right)} \tag{2}$$

where:

A = Cross sectional area of filter (m^2); α = Average specific cake resistance (m/kg);
W = Solids loading (kg/m^3); V = Volume of filtrate (m^3); R = Resistance of filter medium (1/m)

Accordingly, the filtration rate is directly proportional to the pressure drop across the cake and the filtration area, and inversely proportional to the filtrate viscosity and the sum of cake and filter medium resistance.

From the practical point of view, the change of filtration pressure, filtration area, filter medium, filtrate viscosity and solids loading is limited[5]. To increase the filterability of the cake, most practical way, therefore, is to decrease the average specific cake resistance α. This could be accomplished by changing the particle shape, increasing the particles size and narrowing the size distribution of the particles in the cake. The latter two factors can be manipulated by polymer flocculation.

EXPERIMENTAL

Materials

An apatite sample analyzing 26.5% P_2O_5, 43.5% CaO, 1.2% Al_2O_3, 0.8% Fe_2O_3, 0.4% MgO, 0.88% SiO_2 (active), 0.6% Na_2O, 11.8% Insols was used in this study. Superfloc 206 (polyacrylamide, anionic, high charge density, 10-12 million molecular weight), Magnifloc 496C (cationic, high charge density, 10-12 million molecular weight), and some other polymers of the same category but different molecular weights and/or charge densities were used as flocculants. Reagent grade phosphoric acid (85% H_3PO_4) and D.I. Water were used throughout the investigation.

Procedures

17.67g of apatite was dissolved in a solution made up of 270 ml D.I. water and 175 ml 85% H_3PO_4 reagent grade phosphoric acid at 80°C. The slurry was agitated at 400 rpm. 6.6 ml of 98% H_2SO_4 was equally divided into two parts and added into the slurry at two hour intervals. The slurry was agitated for four more hours after the second sulfuric acid addition. This was followed by filtration of the slurry[6].

The polymer was added into the slurry just before filtration and stirred for 1 minute. The filtration was done at 80°C under vacuum (-380 mmHg). The filtration rate was determined by measuring "cake formation time" which is the time elapsed between when the vacuum is applied and the disappearance of the last drop of liquid from the surface of the cake.

Settling rate and sediment volume of gypsum particles were measured to evaluate the size and density of gypsum flocs. The settling rate was measured by observing the descending mud line with time. The sediment volume was determined by sampling 20 ml of the flocculated slurry and pouring it into a 100 ml graduated cylinder filled with 80 ml of product acid. When no further change in the height occurred, the sediment volume was recorded. The settling rate and sediment volume experiments were performed at 80°C.

The filter medium resistance was determined by measuring the time of filtering 350 ml isopropanol at room temperature. The viscosity was monitored by Ubbelohde Viscometer.

RESULTS AND DISCUSSION

First, the effect of polymer dosage on filtration rate of gypsum was investigated. It can be seen from Table I that 5 ppm is the optimum dosage under the present experimental conditions. The cake formation time first decreased then increased slightly as the polymer dosage increased beyond 5 ppm. It is more obvious in the case of cationic polymer where a dramatic increase in cake formation time is noted above 10 ppm dosage. In the case of anionic polymer, no significant effect of polymer molecular weight and charge density was observed.

In order to understand the above observations, the settling rate and sediment volume of gypsum flocs were determined. The filter medium resistance and filtrate viscosity were also measured. It was found that there was no change in viscosity with or without the polymers. The settling rates and sediment volumes are plotted in Figure 1 and Figure 2 respectively. The filter medium resistance data are summarized in Table II.

It can be observed from Figure 1 and 2 that as polymer concentration increases, the settling rate increases and sediment volume decreases. This is indicative of the formation of larger and more deformable flocs at higher polymer dosages.

The filter medium resistance did not change irrespective of anionic polymer dosage (see Table II). However, with Magnifloc 496C, a cationic polymer, the filter medium resistance was not only higher than Superfloc 206, it also increased with the increase in polymer dosage.

It is suspected that amorphous silica, which is expected to be positively charged under the present experimental conditions[7], is flocculated more effectively with oppositely charged anionic polymer (SF 206). The cationic polymer may also flocculate the silica particles, considering that the zeta potential of the particles will be small as a consequence of high ionic strength. Heterocoagulation between positively charged silica and negatively charged gypsum precipitates is expected, but removal of the amorphous silica coating from gypsum particles under intense agitation experienced by the suspension cannot be ruled out. It is therefore, hypothesized that dispersion and flocculation of amorphous silica and fine gypsum precipitates, along with deformability of flocs, governs the filter medium resistance and overall filtration rate.

SUMMARY

The filtration rate of gypsum increased when anionic and cationic polymer flocculants were employed as filter aids. However, anionic polymer was found to be more effective

Table I. The Effect of Polymer Dosage on Gypsum Filtration Rate

	Polymer	Polymer Dosage (ppm) 0	5	10	15	30
Cake Formation Time (Sec)	S.F. 206 (anionic)	325	70	---	125	210
	Magnifloc 496C (cationic)	325	235	395	890	---

Table II. The Effect of Polymer Flocculation on Filter Medium Resistance

Polymer/Dosage (ppm)	Filtration Time of 350 ml Isopropanol, (Sec.)
No Polymer	300
S.F. 206/5ppm	39
S.F. 206/15ppm	35
S.F. 206/30ppm	36
Magnifloc 496C/5ppm	110
Magnifloc 496C/15ppm	274

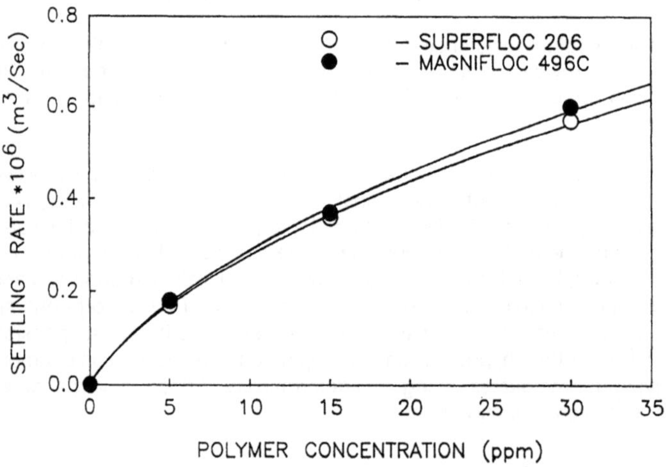

Figure 1. Effect of Polymer Concentration on Settling Rate of Flocculated Gypsum.

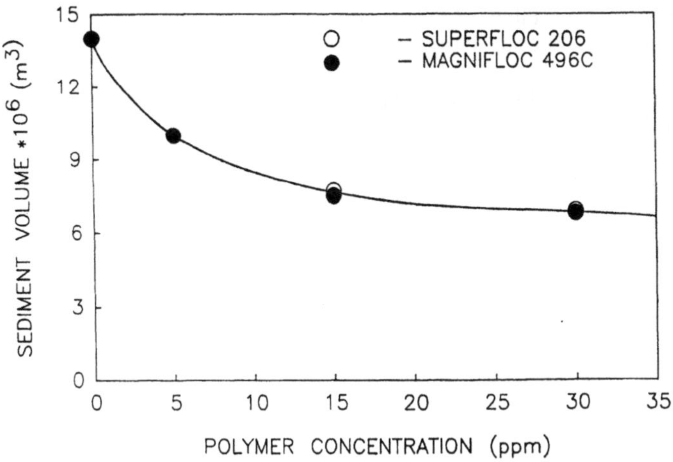

Figure 2. Effect of Polymer Concentration on Sediment Volume of Flocculated Gypsum.

than the cationic flocculant. This can be attributed to the presence of positively charged fine silica particles upon digestion of phosphate rock which can be more effectively flocculated with a oppositely (negatively) charged polymer[7]. An optimum in polymer dosage is explained on the basis of floc deformability and stability of amorphous silica, both increasing beyond a specific flocculant dosage.

ACKNOWLEDGEMENTS

Financial support for this work by NSF-PYI Award (MSM 8821815) and Agrico Chemical Co., Gardinier Inc., and Occidental Chemical Co. is acknowledged.

REFERENCES

1. Maćkiewicz, Jolanta, "The Development of Flocculation Effects in Filter Theory," Chem. Eng. Commun., 23, 1983, pp 305-314.

2. Grace, H.P., "Resistance and Compressibility of Filter Cakes," Chemical Engineering Progress, 49(6), 1953, pp 303-318, 49(7), 1953, pp 367-377.

3. Tiller, F.M. Cooper, H.R., "The Role of Porosity in Filtration V. Constant Pressure Filtration," AIChE Journal, 6(4), 1960, pp 595-601.

4. Ruth, B.F., Montillon, G.H. and Montonna, R.E., "Studies in Filtration," Industrial and Engineering Chemistry, 25(1), 1933 pp 76-82, 25(2), 1933, pp 153-161, 27(6), 1935, pp 708-723.

5. Ruth, B.F. "Correlating Filtration Theory with Industrial Practice," Industrial and Engineering Chemistry, 36(6), 1986, pp 504-571.

6. Taylor, J.F., "The Effect of Polymeric Additives on the Filtration Characteristics of Gypsum," M.S. Thesis, University of Florida, 1989.

7. Becker, P., "Phosphate and Phosphoric Acid," Marcel Dekker, Inc., New York and Basel, 1983.

TRANSITION VELOCITY FOR THE ANNULAR FLOW OF VISCOPLASTIC SUSPENSIONS

Ana E. Paixão
Chemical Engineering Department
Federal University of Sergipe - SE - Brazil

Cesar C. Santana
Chemical Engineering Department
State University of Campinas - SP - Brazil

In this work are obtained generalized Reynolds and Hedstrom numbers connected with a three parameter rheological model to correlate the friction coefficient for the laminar, transitional and turbulent regime in annular flow. The use at experimental data covering a considerable range of dimensionless numbers for the flow of bentonite suspensions leads to a calculation technique for the transition velocity and pressure drop of these suspensions in annular geometries.

INTRODUCTION

Several solid-liquid suspensions frequently encountered in industry exhibit a non-newtonian behavior with an yield stress, a critical value of stress below which they do not flow. These suspensions are extensively used for example in the oil well drilling operations and shows also a non-linear rheological fitting at stresses higher than critical, needing in general the use of more sophisticated rheological models to describe its laminar behavior (Bird et all [1]). Because of the considerable range of flow rates and diameter sizes used in pratice, it is important to master calculations techniques involving also the transitional and turbulent regime of these fluids. This paper is concerned with the annular flow between concentric pipes using dimensionless numbers obtained from the analysis of the analytical solution of fluids following the Robertson and Stiff[2] model.

By using experimental data from a pilot plant constructed to simulate the annular flow of clay suspensions in drilling systems and also date from the literature (Langlinais at all [3]), it was possible to extend the results in order to predict the laminar-turbulent transition velocity and the pressure drop in turbulent flow.

DIMENSIONLESS NUMBERS RELATED TO ANNULAR FLOW

One successful model to the description of shear stress versus shear rate of suspensions is the three-parameter equation proposed by the authors of ref.[2].

$$\tau = A(\dot{\gamma} + C)^B \qquad \tau > \tau_o \tag{1}$$

$$\dot{\gamma} = 0 \qquad \tau \leq \tau_o$$

Where τ_o is the yield stress, being equal to AC^B.

The integration of eq. (1) for the laminar flow on a retangular slot was made by Beirute and Flumerfelt [4]) leading to the relationship between average velocity and pressure drop:

$$\overline{V} = \left(\frac{B}{2B+1}\right)\left[\frac{1}{A}\left(\frac{\Delta P}{L}\right)\right]^{\frac{1}{B}}\left[R_H - \frac{1}{R_H}\lambda^{\frac{2B+1}{B}}\right] - \frac{C}{2R_H}\left(R_H^2 - \lambda^2\right) \tag{2}$$

In eq.(2) the variable R_H is the hydraulic radius obtained from the approximation of an annulus by a slot, where R_H is one forth of the equivalent diameter given by Deq $=\sqrt{2/3}(D_2$ - $D_1)$. The parameter λ is related with the rheological parameters and pressure drop by:

$$\lambda = \frac{AC^B}{\left(\frac{\Delta P}{L}\right)} \tag{3}$$

We can modify the previous equations by introducing the dimensionless parameter ξ_o defined as:

$$\xi_o = \frac{AC^B}{R_H\left(\frac{\Delta P}{L}\right)} \tag{4}$$

In this way eq. (2) accquire the form:

$$\overline{V} = \frac{R_H B}{(2B+1)}\left(\frac{\tau_o}{A\xi_o}\right)^{\frac{1}{B}}\left[1 - \xi_o^{\frac{(2B+1)}{B}} - \left(\frac{2B+1}{2B}\right)\xi_o^{\frac{1}{B}}(1-\xi_o^2)\right] \tag{5}$$

To accomplish a dimensionless formulation from eq. (2), we can use the definition of a friction factor $f = \frac{D.\Delta P/L}{2\rho \overline{V}^2}$ related to the generalized Reynolds number Re^*_{RS} by the equation:

$$f = \frac{24}{\psi Re^*_{RS}} \tag{6}$$

In this way, the analysis leads to the following relationships:

$$Re^*_{RS} = \frac{\rho \overline{V}^{(2-B)} D_{eq}^B}{12^{B-1}\left(\frac{2B+1}{3B}\right)^B A} \tag{7}$$

$$Re^*_{RS} = \frac{1}{12} He^* \left(\frac{3B}{2B+1}\right)\left(\frac{\psi}{\xi_o}\right)^{\frac{2-B}{B}} \tag{8}$$

$$He^* = \frac{\rho D_{eq}^2}{\tau_o}\left(\frac{\tau_o}{A}\right)^{\frac{2}{B}} \tag{9}$$

$$\psi = \left[1 - \xi_o^{\frac{2B+1}{B}} - \left(\frac{2B+1}{2B}\right)\xi_o^{\frac{1}{B}}(1-\xi_o^2)\right]^B \tag{10}$$

Equations (6) and (8) are new definitions of the generalized Reynolds and Hedstrom numbers for the rheological model focused in this work.

From eqs. (5) to (10) it is possible to obtain the pressure drop for the laminar regime of the suspensions that follows the Robertson and Stiff of this model flowing on a given annular geometry. The development parallels that one made by Hanks[5] for the Herschel and Bulkley rheological model.

RESULTS AND DISCUSSION

Pressure Gradient

For the study of the annular frictional pressure gradients including the transitional and turbulent flow we can plot the experimental data for the Fanning friction coeficient f in terms of the generalyzed Reynolds number.

In Table 1 the results are summarized for the properties of clay suspensions and geometry of annuli used in several experiments conducted at two different pilot plants, named here as LSU test facility (Langlinais et all[3]) and PETROBRAS test facility (Silva[6]).

As depicted in figures 1 to 3, the range of variables covered by the experiments permits the analysis of laminar flow and also the transitional and turbulent regions on a considerable variation of the dimensionless parameters Re^*_{RS} and He^*.

Table 1. Description of the Experimental Systems

System nº	ρ (g/cm^3)	Rheological Parameters A (dyn/cm^2)	B (-)	C (s^{-1})	Annular Geometry D_1 (cm)	D_2 (cm)	Pilot Plant (Ref. nº)
1	1.030	0.0713	0.9563	58.39	3.34	6.20	LSU(3)
2	1.054	0.3819	0.8627	49.32	7.30	12.42	LSU (3)
3	1.054	0.2996	0.8210	2.84	3.34	6.20	LSU(3)
4	1.054	0.5592	0.8089	23.87	3.34	6.20	LSU(3)
5	1.054	0.5592	0.8089	23.87	7.30	12.42	LSU(3)
6	1.060	0.5430	0.8178	5.30	3.34	6.20	LSU(3)
7	1.060	0.5430	0.8178	5.30	7.30	12.42	LSU(3)
8	1.220	0.9410	0.6913	31.43	4.82	7.75	PETROBRAS(6)
9	1.220	0.0421	1.0946	752.54	6.10	7.75	PETROBRAS(6)
10	1.190	0.0269	1.5662	176.03	4.82	7.75	PETROBRAS(6)
11	1.000	0.1643	1.1274	27.05	4.82	7.75	PETROBRAS(6)
12	1.000	5.4207	0.7564	28.57	4.82	7.75	PETROBRAS(6)
13	1.000	0.0332	1.2942	241.92	6.10	7.75	PETROBRAS(6)
14	1.000	0.0036	1.5571	305.54	4.82	7.75	PETROBRAS(6)
15	1.000	0.0036	1.5571	305.54	6.10	7.75	PETROBRAS(6)

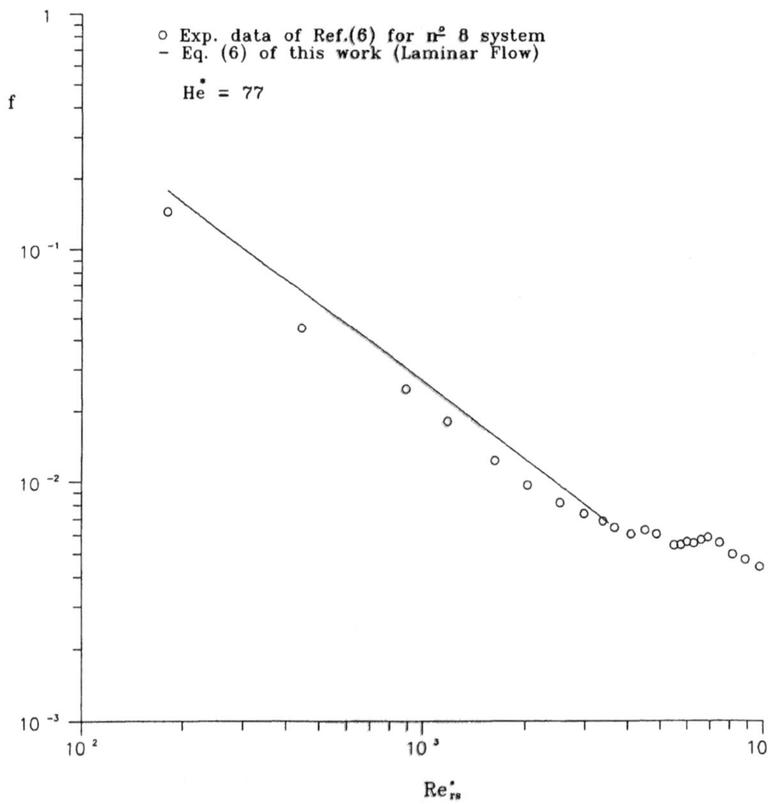

Fig. 1 Friction coefficient as a function of the generalized Reynolds number for number 8 system.

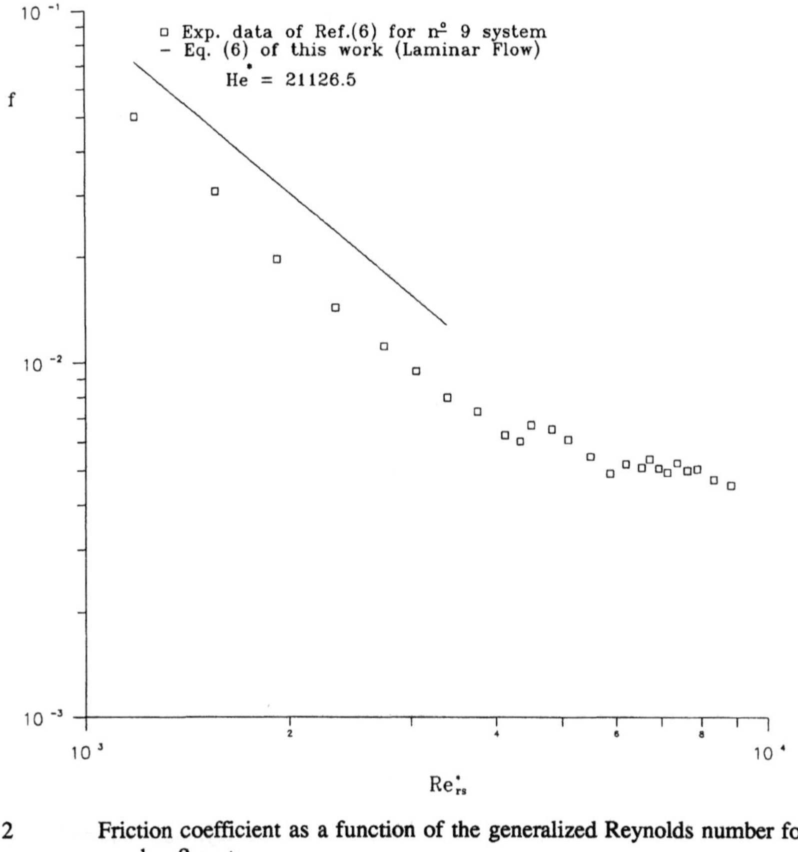

Fig. 2 Friction coefficient as a function of the generalized Reynolds number for the number 9 system.

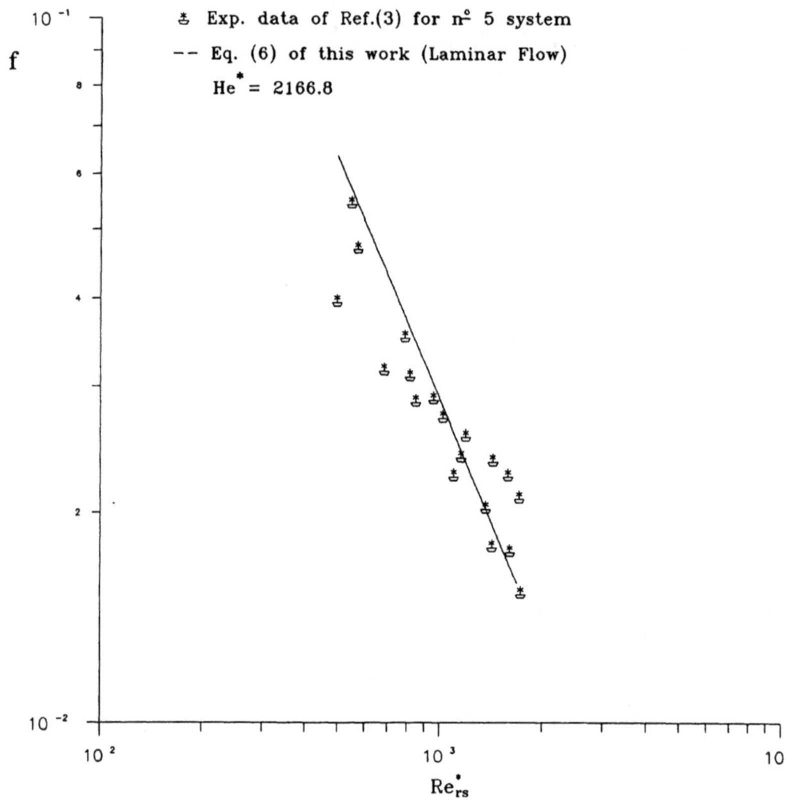

Fig. 3 Friction coefficient as a function of the generalized Reynolds number for the number 5 system.

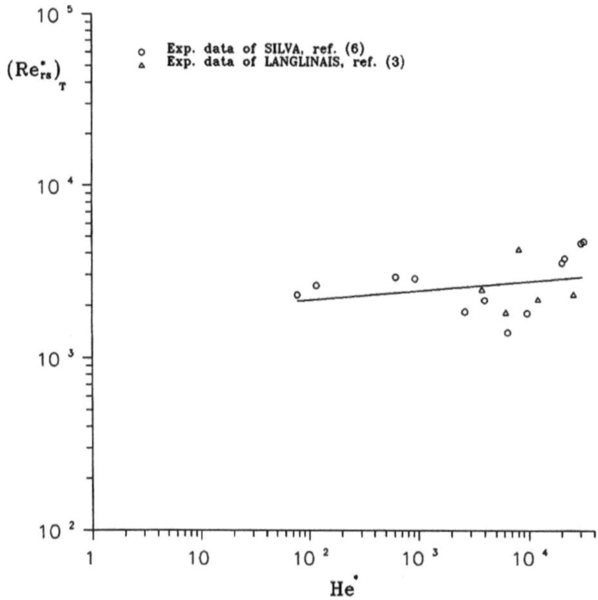

Fig. 4 Dimensionless plot to predict transitional velocity.

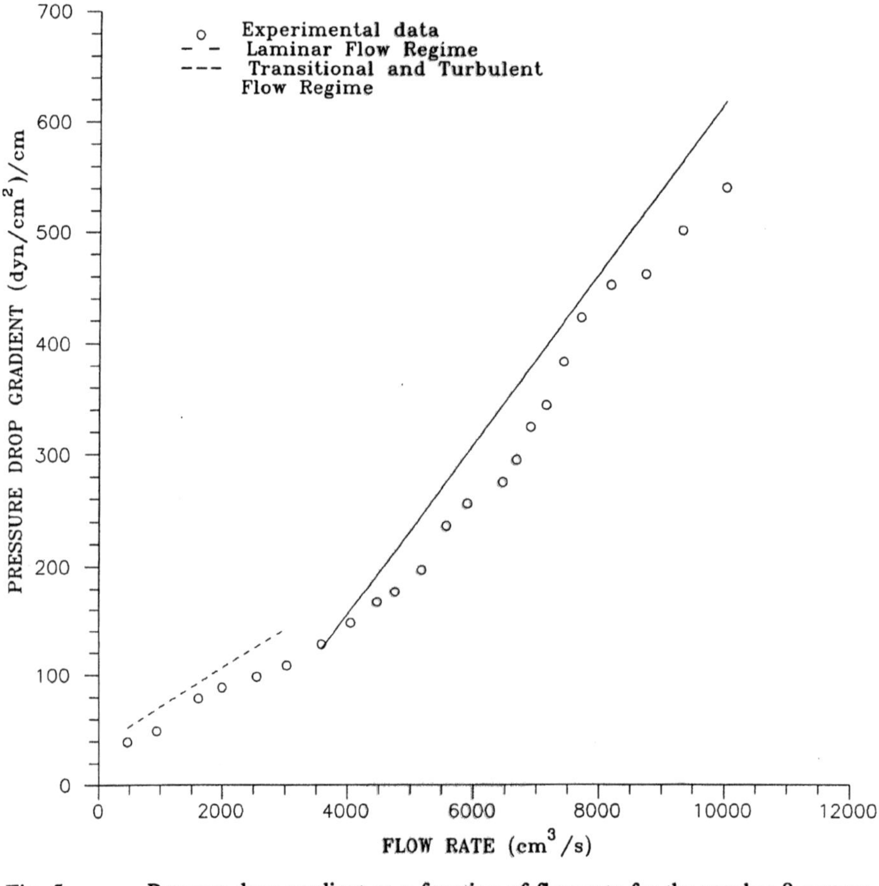

Fig. 5 Pressure drop gradient as a function of flow rate for the number 8 system.

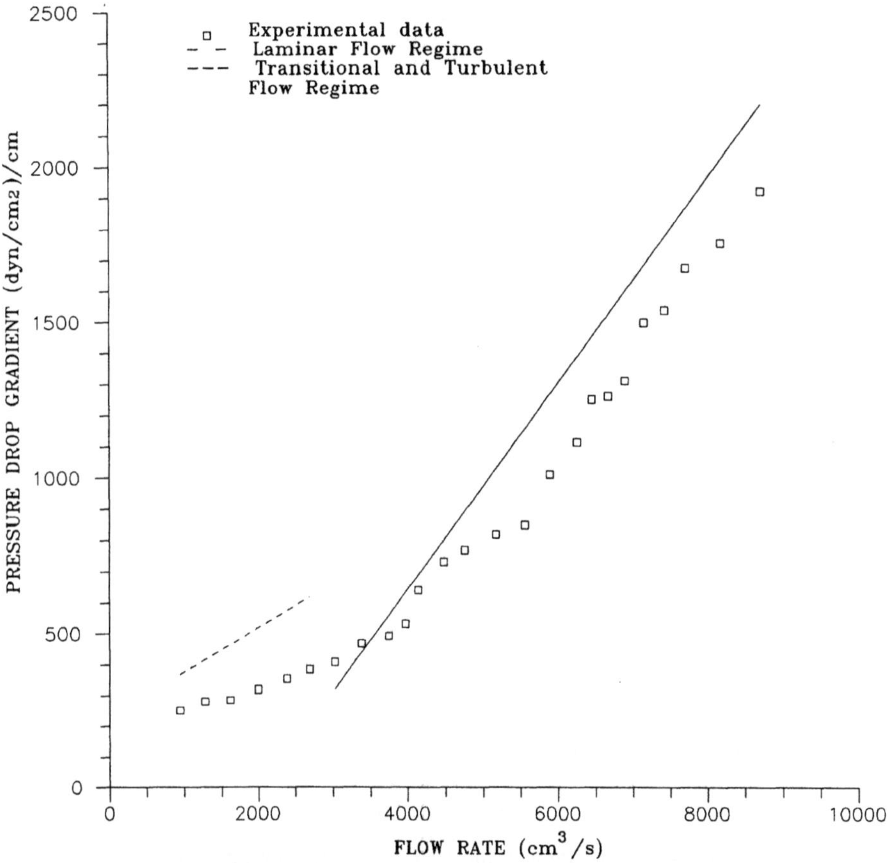

Fig. 6 Pressure drop gradient as a function of flow rate for the number 9 system.

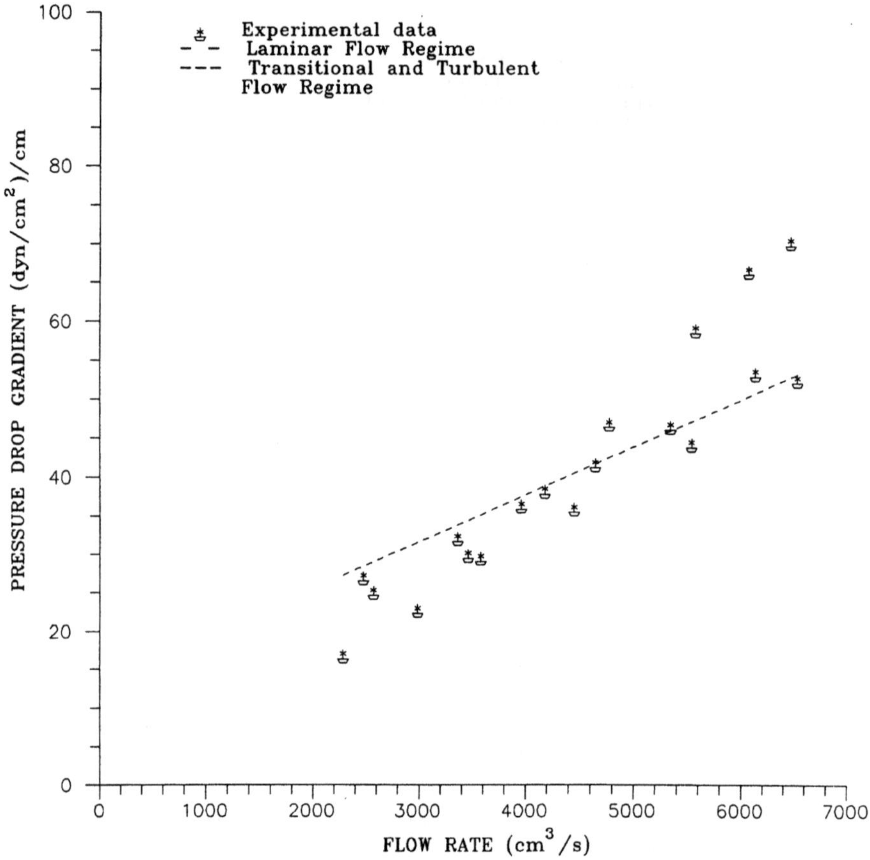

Fig. 7 Pressure drop gradient as a function of flow rate for the number 5 system.

Transitional Velocity Determination

The observation of the departure of experimental points from the straight line characteristic of the laminar flow can determine a criterion to ascertain the transition Reynolds number. As the turbulent region points can also be fitted by a straight line, the critical Re^*_{RS} is obtained here at the interception of both straight lines for each value of He^*.

An empirical correlation giving the relationship between the transitional Re^*_{RS} and He^* is shown in Fig. 4, and can be used as procedure for the determination of this important dynamic variable at the onset of turbulence. The numerical adjustment for the function Re^*_{RS} at the transition is:

$$(Re^*_{RS})_T = 1679{,}6\,(He^*)^{0{,}054} \tag{11}$$

Although the analytical procedure developed previously 2 is recognized as as a sound formulation for the laminar regime, at this point of the work development the turbulent region can only be described by an empirical fitting based on a Blasius form of relation hip between f and Re^*_{RS}. The set of experimental data considered in this work furnishes the following correlation:

$$f = \frac{0.221}{(Re^*_{RS})^{0.408}} \tag{12}$$

A comparison between experimental and calculated pressure drops for the entire range of flow rates is showed in figures 5 to 7.

CONCLUSIONS

The utilization of a three-parameter rheological model to describe the rheology of clay suspensions conducted to a development of new dimensionless parameters named here as generalized Reynolds and Hedstrom numbers permits the analytical relationship between pressure drop and flow rates for the laminar flow in annular flow.

The semi-empirical analysis of many experimental data collected in two different pilot plants led to satisfactory correlations for the turbulent friction coefficient and to the prediction of laminar-turbulent transition velocity for the flow of viscoplastic suspensions in annuli.

Acknowledgements

The authors are indebted to Dr. Julius P. Langlinais from Louisiania State University and to Eng. Maria das Graças Silva from Petrobrás for the supply of experimental data and discussions about annular non-newtonian flow.

REFERENCES

1. Bird, R.B., Dai, G.C. and Yarusso, B.J., The Rheology and Flow of Viscoplastic Materials, Reviews in Chemical Engineering, vol. 1, nº, pp. 1 - 69, (1982).
2. Robertson, R.E. and Stiff, H.A., An Improved Mathematical Model for Relating Shear Stress to Shear Rate in Drilling Fluids and Cement Slurries, Society of Petroleum Engineers Journal, pp. 31 – 36, February, (1976).
3. Langlinais, J.P., Bourgoyne Jr., A.T. and Holden, W.R., Friction Pressure Losses for the Flow of Drilling Mud and Mud/Gas Mixtures, Paper SPE, nº 11993, October, (1983).
4. Beirute, R.M. and Flumerfelt, R.W., An Evaluation of the Robertson - Stiff Model Describing Rheological Properties of Drilling Fluids and Cement Slurries, SPE Journal, pp. 97 - 100, April, (1977).
5. Hanks, R.W., Low Reynolds Number Turbulent Pipeline Flow of Pseudohomogeneous Slurries, Hidrotransport 5, Paper c 2, pp. c 2 - 23 to c 2 - 34, (1978).
6. Silva, M.G.P., Evalution of the Flow Regime of Drilling Fluids in Annular Sections. (In Portuguese), 2nd Drilling Fluids Seminar, Salvador, BA, pp. 1 - 10, (1989).

FLOTATION SEPARATION OF APATITE FROM DOLOMITE USING DODECYLAMINE AND SODIUM CHLORIDE

Brij M. Moudgil and Dursun E. Ince*

Mineral Resources Research Center
Department of Materials Science and Engineering
University of Florida
Gainesville, FL 32611

ABSTRACT

It is critical to separate dolomitic limestone impurities from apatite to utilize the South Florida Phosphate Rock deposits for fertilizer manufacturing. Conventional flotation scheme is not successful for separation of dolomite from its mixture with apatite. Several processing methods utilizing both anionic and cationic collectors have been proposed to solve this problem, but they have yet to be tested on a commercial scale.

Adsorption of the collectors such as dodecylamine and fatty acids is reported to be strongly influenced by electrical double layer forces. In this study, manipulation of the electrostatic forces by adding sodium chloride was attempted to achieve the desired separation. Suitable separation of apatite from dolomite was achieved with dodecylamine hydrochloride in the presence of sodium chloride. Zeta potential measurements revealed that sodium chloride acts as an indifferent electrolyte for dolomite, but causes a shift in the iso-electric point of apatite.

INTRODUCTION

Separation of dolomitic limestone impurities from apatite has become increasingly important since depletion of current high grade phosphate deposits in Florida is fast approaching. Economy of chemical processing of the phosphate rock to phosphoric acid dictates that the phosphate feed to the chemical plants contain no more than 1% MgO. The presence of higher MgO content, among other problems, is reported also to increase the viscosity of the phosphoric acid resulting in higher pumping costs. Most of the Mg exists as discreet dolomite particles and is amenable to removal by flotation processes.

Several attempts in the past have been made to remove dolomite from apatite by flotation, however, process economics for most of the suggested schemes at the plant scale has not yet been established[1-10].

---

* Current Address: Mineral Technology, Pfizer Inc., Easton, PA.

Particle Technology and Surface Phenomena in Minerals and Petroleum
Edited by M.K. Sharma and G.D. Sharma, Plenum Press, New York, 1991

Considering that electrostatic forces alone or in combination with chemisorption forces are responsible for adsorption of phosphate collectors such as sodium oleate and dodecylamine on apatite and dolomite, it was decided to manipulate electrostatic forces by adding inorganic salts such as sodium chloride[11]. It is to be noted that sodium chloride is generally used as an indifferent electrolyte in oxide mineral systems.

EXPERIMENTAL

A sample of high grade phosphate rock (apatite) was obtained from Agrico Chemical Co. and dolomite sample was procured from IMC Fertilizers, Inc. Flotation tests employed 60 x 100 mesh fraction of the sample and electrokinetic and FT-IR studies were conducted on -325 mesh fraction of the ground sample.

For flotation studies, dodecylamine hydrochloride from Eastman Kodak Co. and purified sodium oleate from Fisher Scientific Co. were used as the cationic and anionic collectors, respectively. All other chemicals were reagent grade purchased from Fisher Scientific Co. HNO_3 and KOH were employed as pH modifiers. Deionized water (D.I.) of less than 1.2 micromhos conductivity was used in this study.

Flotation experiments were conducted using a modified Hallimond Cell. Zeta potential measurements were conducted using a Pen Kem Model 501 Laser Zee Meter.

RESULTS AND DISCUSSION

Flotation with Cationic Collector

Single mineral apatite and dolomite flotation tests using dodecylamine hydrochloride as the collector indicated that apatite can be recovered from its mixture with dolomite at pH less than 4.5 at a collector concentration of 1 m mol/m^3, and at pH 9.8 at 0.16 m mol/m^3. However, the magnitude of selectivity predicted by the single mineral test was not realized when attempts were made to float a 50:50 mixture of apatite and dolomite.

Flotation with Anionic Collector

Selective flotation of apatite from dolomite was predicted by single mineral tests in the pH range of 7 to 10, and of dolomite from apatite at pH of about 5, with 4×10^{-5} $kmol/m^3$ of sodium oleate collector. Once again, mixed mineral (50:50) tests did not yield the selectivity predicted on the basis of the single mineral flotation studies.

Moudgil and co-workers[7, 11] and Soto and Iwasaki[10] have attributed the loss of selectivity to electrostatic forces encountered in the above mineral-collector systems. It was therefore, decided to modify the flotation behavior of apatite and dolomite by adding salts such as sodium chloride. In this paper, flotation results with dodecylamine hydrochloride collector are discussed. Flotation studies with sodium oleate collector have been discussed elsewhere[12].

Effect of Sodium Chloride

Single Mineral Flotation Tests. Experiments conducted in the natural pH range (pH 6.7) as shown in Table I, yielded 88% flotation recovery of apatite at a dodecylamine concentration of 1.6×10^{-4} $kmol/m^3$. Apatite recovery in the absence of NaCl was observed to be 49%. In contrast, dolomite recovery under similar experimental conditions with and without NaCl remained at the 18-19% level.

Table I. Effect of Sodium Chloride on The Single Mineral Flotation of Apatite and Dolomite

Mineral	Amount Floated, Weight %	
	Without Sodium Chloride	With Sodium Chloride
Apatite	49.0	87.7
Dolomite	18.5	18.0

Collector: Dodecylamine Hydrochloride, 1.6×10^{-4} kmol/m^3
NaCl Conc: 5.0×10^{-1} kmol/m^3
Feed: 1g, 65 X 100 Mesh Size Fraction
Flotation pH: 6.7

Table II. Results of Mixed Mineral Flotation in the Presence of Sodium Chloride

Material	Chemical Analysis, %		Apatite Recovery, %
	P_2O_5	MgO	
Feed	18.0	9.5	100
Concentrate	33.6	1.0	80

Collector: Dodecylamine Hydrochloride, 1.6×10^{-4} kmol/m^3
NaCl Conc: 5.0×10^{-1} kmol/m^3
Feed: 1g, 88/12 Apatite-Dolomite Mixture
Flotation pH: 6.3

Mixed Mineral Flotation Tests. Flotation of 88:12 apatite-dolomite mixtures was conducted to verify the selectivity predicted on the basis of single mineral experiments. Results presented in Table II demonstrate that at dodecylamine concentration of 1.6×10^{-4} kmol/m^3 apatite can be selectively recovered from the mixture leaving dolomite in the sink fraction.

Zeta Potential Studies

It is clear from the above data that addition of sodium chloride results in selective flotation of apatite from its mixture with dolomite using dodecylamine as the collector at pH 6.3. Single mineral flotation results at pH 6.7 indicated that more apatite floats in the presence of salt than without it. On the other hand, addition of sodium chloride had no significant effect on dolomite flotation,.

Zeta potential measurements of apatite and dolomite with and without sodium chloride are plotted in Figures 1 and 2 respectively. Zeta potential of apatite in the presence of sodium chloride is negative over the entire pH range (4 to 11) examined in this study, indicating reversal of the surface charge below the iso-electric point (iep) of pH 5.4. The zeta potential in the pH range of 6 to 9 was, however, found to be less negative than in distilled water. These measurements indicate that sodium chloride does not act as an indifferent electrolyte for apatite surface.

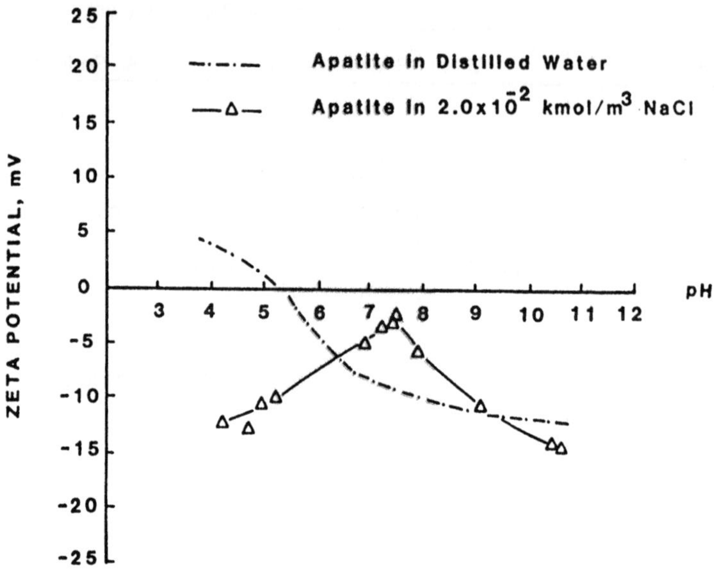

Figure 1. Zeta Potential of Apatite With and Without Sodium Chloride

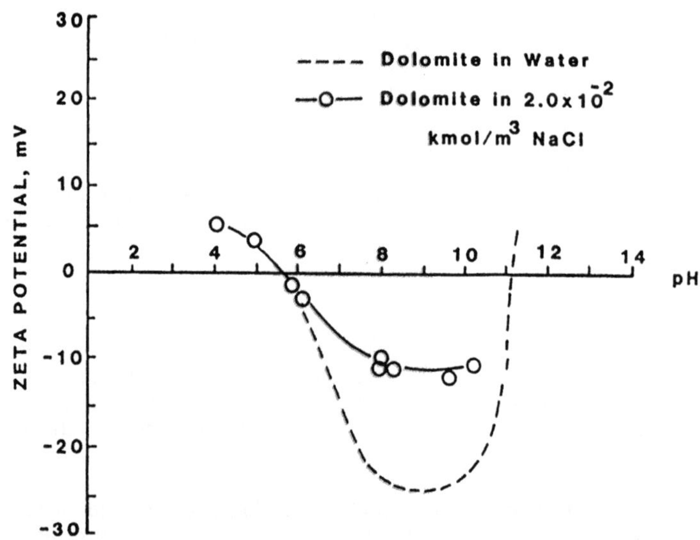

Figure 2. Zeta Potential of Dolomite With and Without Sodium Chloride.

Zeta potential of dolomite in the presence of sodium chloride did not exhibit any shift in the iep, and remained unchanged from deionized water also below the iep (pH 5.3). Above the iep, however, presence of sodium ions decreased the magnitude of zeta potential values. This can be attributed to compression of the electrical double layer due to increased ionic strength as a result of 2 x 10^{-2} kmol/m^3 NaCl addition.

Solution Properties of Dodecylamine Hydrochloride

Using the thermodynamic data of Ananthapadmanabhan[13], species distribution diagram of dodecylamine was constructed (see Figure 3). The maximum in amine-ammonium complex formation at 1.6 X 10^{-4} kmol/m^3 concentration occurs at pH 10. Ananthapadmanabhan[13] and Pugh[14] have shown the significance of such complexes in flotation, even if present in small amounts.

Flotation Behavior of Apatite and Dolomite in the Presence of Sodium Chloride

Flotation of Apatite below its iep (pH 5.4) with dodecylamine has been attributed to chemical interaction between the collector ion and the mineral surface[10]. At pH 6-7, amine monomers (RNH_3^+) and dimers $(RNH_3)_2^{++}$ become dominant species which can adsorb on the negatively charged apatite surface through electrostatic attraction. In the presence of sodium chloride, the magnitude of negative zeta potential decreases, but the flotation is enhanced. The reason for higher apatite flotation when sodium chloride is added is not clear at the present time.

Dolomite flotation below its iep (pH 5.3) is about 20% indicating the presence of weak specific interactions in addition to the coulombic attraction. The heat of reaction of dodecylamine with phosphate and carbonate anions (-15.1 and -3.3 kj/mol for apatite and dolomite, respectively) determined by Soto and Iwasaki[10], supports this hypothesis. The heat of reaction measurements indicate that the collector would adsorb preferentially on apatite than dolomite when both the minerals are present. Addition of sodium chloride decreases the magnitude of negative zeta potential, thus reducing the adsorption of the cationic collector and enhancing the selectivity in the mixed mineral flotation.

CONCLUSIONS

Separation of apatite from dolomite using dodecylamine as the collector is enhanced by sodium chloride addition. Adsorption of the cationic collector on the apatite surface is attributed mostly to chemical interactions. Sodium chloride, as indicated by zeta potential measurements, does not act as an indifferent electrolyte and affects the surface structure of apatite. However, in the case of dolomite, sodium chloride reduces the adsorption of the collector by compressing the electrical double layer. Satisfactory separation of apatite from its mixture with dolomite is achieved at pH 6.3 with dodecylamine collector in the presence of sodium chloride.

ACKNOWLEDGEMENTS

Financial support for this work by the Florida Institute of Phosphate Research (Grant # 85-02-067) is acknowledged. Any opinions, findings, and conclusions or recommendations expressed in this work are those of the authors and do not necessarily reflect the views of the Florida Institute of Phosphate Research.

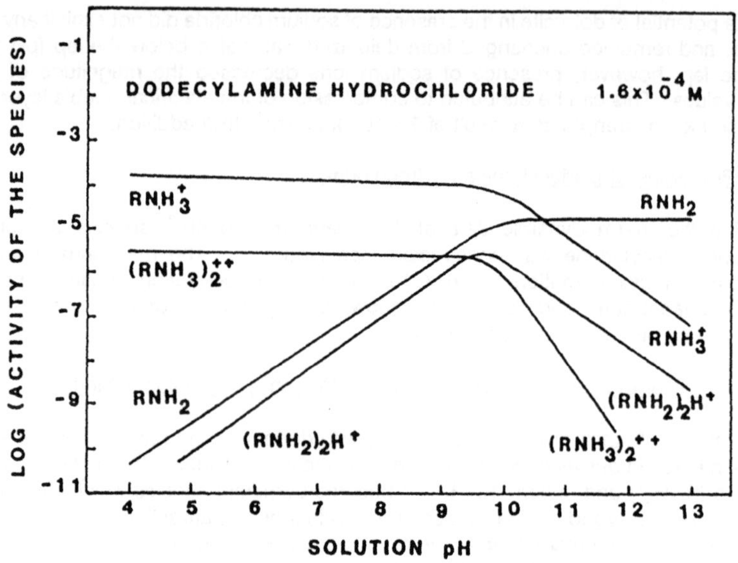

Figure 3. Dodecylamine Species Distribution as a Function of pH.
Total Amine Concentration, 1.6×10^{-2} kmol/m^3.

REFERENCES

1. Johnston, D.J. and Leja, J., "Flotation Behavior of Calcium Phosphates and Carbonates in Orthophosphate Solution," IMM Trans, Vol. 87, pp. C237-242, 1978.

2. Lawver, J.E., Bernardi, J.P., McKereghan, G.F., Raulerson, J.D., Lynch, D. and Hearon, R.S., "New techniques in Beneficiation of the Florida Phosphates of the Future," Minerals and Metallurgical Processing, Vol. 1, No. 2, pp. 89-106, 1984.

3. Lawver, J.E., Raulerson, J.D. and Cook, C.C., "New Techniques in Beneficiation of Phosphate Rock," Trans., SME/AIME, Vol. 268, pp. 1787-1801, 1980.

4. Lawver, J.E., Wiegel, R.L., Snow, R.E. and Hwang, C.L., "Beneficiation of Dolomitic Florida Phosphate Reserves," XIV International Mineral Processing Congress, Toronto, Canada, 1982.

5. Lehr, J.R. and Hsieh, S.S., "Beneficiation of High Carbonate Phosphate Ores," U.S. Patent 4,287,053, 1981.

6. Llewellyn, T.O., Davis, B.E. and Sullivan, G.E. "Beneficiation of Florida Dolomite Phosphate Ores," Minerals and Metallurgical Processing, Vol. 1, No. 1, pp. 43-48, 1984.

7. Moudgil, B.M. and Chanchani, R., "Selective Flotation of Dolomite from Francolite Using Two-Stage Conditioning," Minerals and Metallurgical Processing, Vol. 2, No. 1, pp. 19-25, 1985.

8. Moudgil, B.M. and Somasundaran, P., "Advances in Phosphate Flotation," In: Advances in Mineral Processing, P. Somasundaran, Ed., SME/AIME Pub., Littleton, Colorado, pp. 426-441, 1986.

9. Snow, R.E., "Flotation of Phosphate Ores Containing Dolomite," U.S. Patent No. 4,364,824.

10. Soto, H. and Iwasaki, I., "Flotation of Apatite from Calcareous Ores with Primary Amines," Minerals and Metallurgical Processing, Vol. 2, pp. 160-166, 1985.

11. Moudgil, B.M., and Ince, D.E., "Method of Beneficiating Phosphate Ores," U.S. Patents 4,814,069 and 4,857,174, 1989.

12. Ince, D.E., "Effect of Sodium Chloride on the Selective Flotation of Dolomite from Apatite", Ph.D. Dissertation, University of Florida, Gainesville, FL, 1987.

13. Ananthapadmanabhan, K.P, "Association Interactions in Surfactant Solution and Their Role in Flotation," Doctor of Engineering Science Dissertation, Columbia University, New York, NY, 1980.

14. Pugh, R.J., "The Role of the Solution Chemistry of Dodecylamine and Oleic Acid Collectors in the Flotation of Fluorite," Colloids and Surfaces, Vol. 18 pp. 19-41, 1986.

SURFACTANTS IN ENHANCED PETROLEUM RECOVERY PROCESSES: AN OVERVIEW

Mahendra K. Sharma

Research Laboratories
Eastman Chemical company
Kingsport, TN 37662 (USA)

This paper reviews various aspects of surfactant flooding for enhanced petroleum recovery. Surfactants have been suggested and tested in several forms such as foams, macroemulsions and microemulsions to recover additional oil from petroleum reservoirs which is unrecoverable by conventional methods. Several important parameters such as optimal salinity, surfactant adsorption, surfactant/polymer interactions, ultralow interfacial tension, interfacial viscosity, emulsion and foam stability, foaminess, bubble size and size distribution, surface charge, and propagation of oil ganglia were described in relation to oil displacement in porous media. For the foam flooding process, the effect of polymer flooding on oil recovery was reported. Results indicated that the surface properties of the foaming agents significantly influenced the oil displacement, whereas bulk properties of the fluid in porous media did not influence the oil recovery. For microemulsion flooding, the optimal salinity of the system can be shifted to a desired value by varying the concentration and structure of the surfactant and cosurfactant. The surfactant formulations composed of petroleum sulfonates and ethoxylated sulfonates are relatively insensitive to divalent cations. Based on the experimental findings, a maximum in oil recovery was observed when a minimum in interfacial tension, apparent viscosity, surfactant loss and coalescence rate of oil droplets occurred at an optimal salinity.

INTRODUCTION

In past decades much emphasis has been given to enhanced petroleum recovery by several flooding processes[1-15] in order to meet the increasing demand of the world-wide energy consumption. Recovery processes can be divided into four major processes: chemical, thermal, miscible and microbial flooding (Figure 1). Among the various processes of petroleum recovery, thermal process offers a promising approach to recovering about 70% of the world's enhanced oil recovery (EOR) production. Several different types of chemicals have been employed for petroleum recovery. Among these chemicals, surfactants were found to be the most promising in EOR processes.[16-25]

Surfactant molecules are composed of two portions known as hydrophilic and hydrophobic groups. The hydrophilic group has sufficient solubility in an aqueous phase and always tends to bring the entire molecule into solution, while the hydrophobic group is insoluble in an aqueous phase and tends to expel the entire molecule from the solution. If the force due to the hydrophobic group predominates, the surfactant molecules tend to concentrate at an interface and at least a part of the hydrophobic group is not in contact with molecules of the aqueous phase. The hydrophobic group is usually composed of a long straight or branched hydrocarbon chain. A surfactant molecule is schematically illustrated in Figure 2. The surface properties of a surfactant molecule are controlled by the balance between its hydrophilic and hydrophobic characteristics. If a surfactant molecule is composed of a hydrocarbon chain with less than 12 carbon atoms, it is generally water soluble because the polar head group drags the entire molecule in water. If the hydrocarbon chain length is greater than 14 carbon atoms, the surfactant is water insoluble.

Surfactant molecules in solution beyond their critical micelle concentration (CMC) are widely known to form aggregates in different shapes. Below the CMC, surfactants in solution are present in the form of individual molecules. Figure 3 illustrates the formation of various association structures with increasing surfactant concentration. It is likely that surfactant molecules may form spherical, cylindrical, hexagonal, lamellar and reversed micellar (e.g. spherical) structures in solution by adjusting the proper physicochemical conditions such as pH, temperature and the presence of various electrolytes. If oil is present in the system, these association structures can solubilize the oil, and can produce a clear, thermodynamically stable system. Depending on the nature of the oil phase and the oil/water ratio, the oil can be a continuous or disperse phase in the system.

The molecular structure of surfactants controls not only the concentration of the surfactants at the interface and the resulting reduction in surface/interfacial tensions, but also affects the orientation of the molecules at the interface. The hydrophilic group is either ionic in nature or highly polar. Based on the nature of the polar group, surfactants can be classified as anionic, cationic, non-ionic or amphoteric. Among these types, anionic and non-ionic surfactants are preferably employed in enhanced oil recovery processes (EOR) due to their low adsorption on reservoir rocks. Therefore, these surfactants are briefly described.

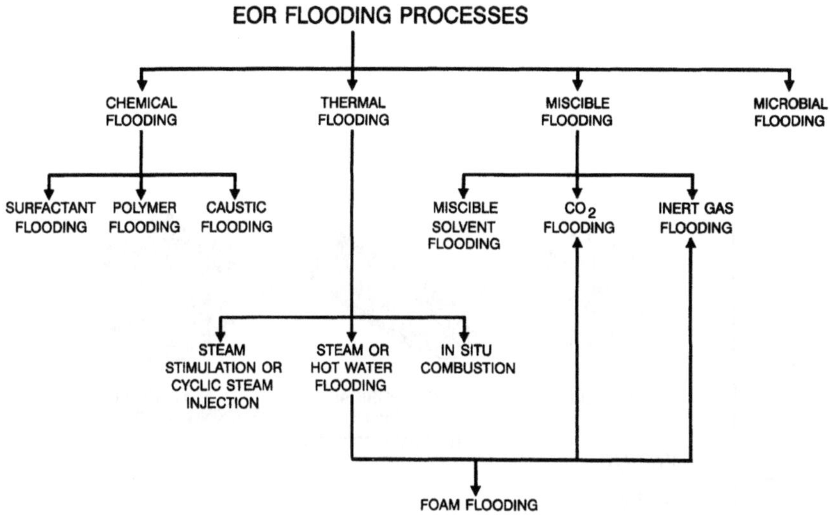

Figure 1. A Schematic Illustration of Various Oil Recovery Processes.

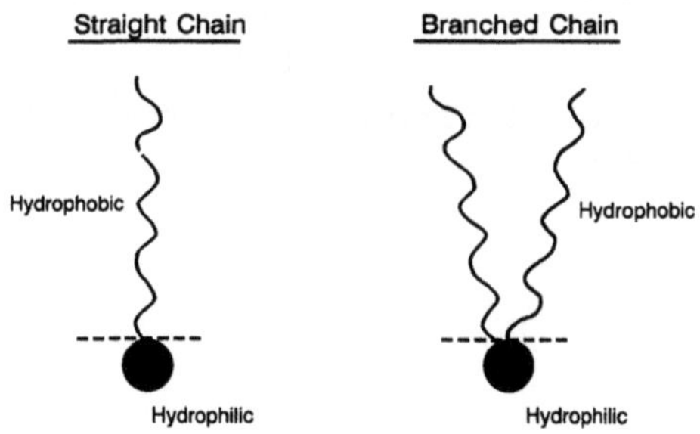

Figure 2. Possible Structures of the Surfactants.

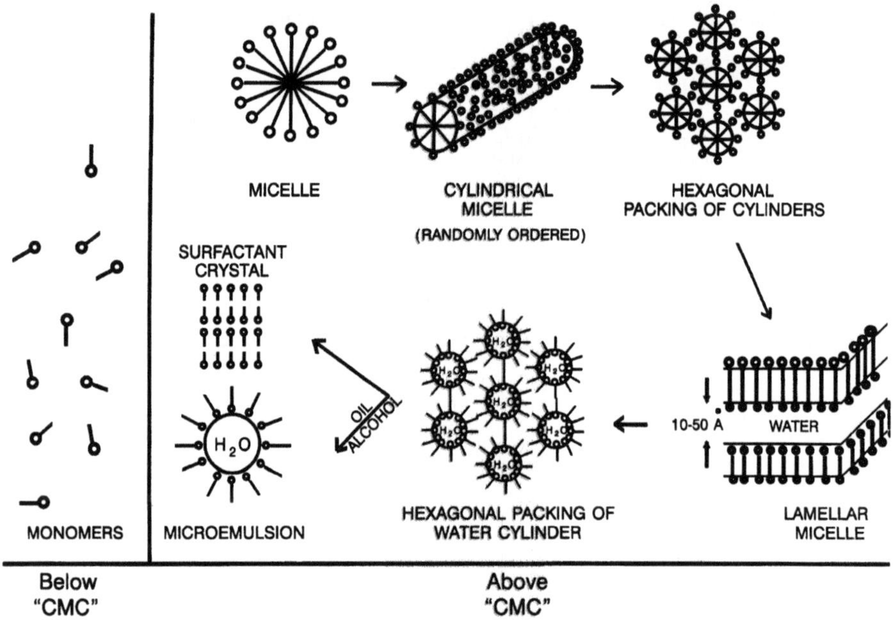

Figure 3. A Schematic Illustration of Surfactant Association Structures.

EOR SURFACTANTS

Anionic surfactants with sulfonates, carboxylates, sulfate and posphonate monoester groups as well as biosurfactants have been employed in oil recovery processes. Among these groups, sulfonates have been widely used in EOR processes during the past years.[26,27]

Petroleum Sulfonates: Most of the surfactant formulations for EOR involve petroleum sulfonates. These petroleum sulfonates are prepared from selected refinery products. Among several refinery products, the furfural extract from solvent lube operations is considered an attactive feedstock for oil recovery sulfonates. Lube oil stocks are sulfonated to separate aromatic components. The petroleum sulfonate is in short supply due to the limited demand for lube oil stocks.

The petroleum sulfonates were also prepared by sulfonating gas/oil fractions with sulfuric acid in the 1970's. The main disadvantage of this process was sluge disposal. Therefore, the gas/oil fractions were sulfonated with sulfur trioxide in order to eliminate the sluge disposal problem. The process was not cost-effective. Thus, crude oil was selected to sulfonate with sulfur trioxide. A few years later, Marathon Oil Company manufactured petroleum sulfonates by a crude oil sulfonation process. These petroleum sulfonates have been employed in several EOR field projects in the United States.

Lignosulfonates: Lignosulfonates are wood-based chemicals made from the sulfite liquors generated during the sulfite process of wood pulping. These are cosiderably cheaper than the well-known petroleum sulfonates, which are currently popular and widely employed in many micellar recovery strategies for oil displacement in porous media. The lignosulfonates are water soluble and anionic in nature. Their properties can be modified by adjusting sulfonate content, molecular weight, type and amount of counter ions, and surface activity. Several lignosulfonate products known as Marasperse are commercially available from American Can Company. The basic functional constituent of these products is lignosulfonates, an anionic polyeletrolyte of molecular weight between 1,000 to 20,000.

Marasperse lignosulfonates can produce stable emulsions. These emulsions are resistant to pH and eletrolyte contents. It has been observed that sodium hydroxide enhances the surface activity of aqueous lignosulfonate solutions, which in turn lowers the interfacial tension at the oil/water interface, thereby rendering the emulsions more stable. The emulsions stabilized by lignosulfonates are also resistant to mechanical agitations and to large temperature variations.

The emulsions stabilized by lignosulfonates are extremely stable under a wide range of condition. Therefore, the conventional emulsion breaking techniques are not very effective to separate oil and aqueous phases. The most effective method to break lignosulfonate-stabilized emulsions is by incorporating a cationic surface active agent. Several articles have been published on the emulsification and oil recovery by lignosulfonates.[28-32]

<u>Synthetic Sulfonates</u>: The synthetic sulfonates are reported to be more effective in displacing oil, but are more expensive as compared to petroleum sulfonates. These sulfonates also have better tolerance to metal ions in solution such as Ca^{++} and Mg^{++}. The synthetic sulfonates were introduced in World War II, but these were not available commercially until the late 1940's. The alkyl benzene sulfonates were the first commercially available important synthetic sulfonate. The aromatic and olefin feedstocks have been identified as the major sources for alkylate preparation. These alkylates are sulfonated with either sulfur trioxide or sulfuric acid to prepare synthetic sulfonates.

A process to manufacture synthetic sulfonates was developed by Suntech Inc.. According to this process, caustic is used to neutralize the product, which minimizes the water usage during the process. Various companies such as Exxon and Shell have employed synthetic sulfonates for improving oil recovery. Several more expensive synthetic sulfonates also have been used as additives in the EOR processes. In such cases, a portion of the petroleum sulfonate prepared by sulfonation of gas/oil or crude oil is replaced with the synthetic sulfonates. It has been observed that the surfactant formulations containing mixed petroleum sulfonate and synthetic sulfonate have resulted in improved oil recovery.

<u>Ethoxylated Sulfonates</u>: The condensation of fatty alcohols and ethylene oxide yields non-ionic sufactants, which are low emulsifiers and low foamers. The ethoxylated sulfonates can be prepared by sulfonating ethoxylated alcohols. Sulfonated alcohols are excellent foamers as are the sulfonated versions of non-ionic surfactants. The ethoxylated sulfonates are also known as alcohol ether sulfonates. The laboratory results and field tests indicate that the mixed surfactant formulation composed of a petroleum sulfonate and ethoxylated sulfonate or alcohol can improve oil recovery as well as increase salinity tolerance. Ethoxylated sulfonates also lead to a greater flexibility in fluid system design because the degree of ethoxylation and/or the chain length may be varied to adjust slug viscosity.

Previous investigators[20] have shown that by replacing petroleum sulfonates with the ethoxylated sulfonate EOR-200, the optimal salinity can be increased to as high as 32% NaCl brine (Table 1). These results demonstrated that the mixed petroleum sulfonates and ethoxylated sulfonates or alcohol formulations are promising candidates for reservoirs containing formation water of high salinity. Recently, Texaco used an ethoxylated-alcohol as a cosurfactant in EOR formulations. These formulations have resulted in improved oil recovery, and have also allowed slugs to be tailored for higher temperature and salinity of the reservoir.

Table 1. Optimal salinity of surfactant formulation containing TRS 10-410, EOR-200 and Isobutyl Alcohol.

TRS 10-410 (wt%)	EOR-200 (wt%)	OPTIMAL SALINITY (% NaCl)
5.0	0.0	1.9
4.0	1.0	5.8
3.0	2.0	11.0
2.0	3.0	21.3
1.0	4.0	31.7

Ethoxylated Carboxylates: During the past years, the use of ethoxylated carboxylates in EOR formulations have been ignored by the investigators due to the fact that the salt to free acid ratio is too dependent on the pH of the system. It was believed that these complex formulations would lead to inexplicable phase behavior, especially under conditions of the petroleum reservoir. Recently, interest has developed in investigating the phase behavior of ethoxylated carboxylates with the hope that the electronegative oxygen, beta to the carboxylate group, will minimize the pKa value of the acid, which in turn reduce the pH sensitivity for their applications in the petroleum recovery processes.[33]

It is reported that the ethoxylated carboxalates have the tendency to form liquid crystals at low alkane carbon number (ACN), high pH and low temperature. A pH increase causes the ionization degree of the surfactant molecules to increase, which increases water solubility. Therefore, the salinity required to form a middle-phase microemulsion increases. In addition, an increase in electrolyte concentration results in an increase in micellar aggregation number, which leads to an increased molecular packaging with the resulting formation of liquid crystals. Several attempts were made to eliminate the formation of liquid crystals in the system by employing short chain alcohols as cosolvents. The solubilization parameters were found to be lower for ethoxylated carboxylate containing systems when compared to systems without alcohol. At the present time, there is no sufficient experimental evidence to explain these complex systems, but increased interest would certainly allow exploring the utility of these systems in the EOR processes in the coming years.

Biosurfactants: Biosurfactants are surface active compounds derived from biological sources which, like synthetic surfactants, exhibit characteristic physical and chemical properties. The production of suface active materials by microorganisms has been of general long-standing recognition with a systematic characterization of such products slow to emerge.

The biosurfactants or microbial surface active agents,[34,35] which have been discussed, involve surfactants, stabilizers, flocculating agents, foaming agents, emulsifiers, and de-emulsifiers. Most of these materials are poorly characterized. Biosurfactants, which have been characterized so far, are lipids. The lipid molecules usually have the classical surfactant structure. Like surfactants most lipid molecules consist of a hydrophobic hydrocarbon chain and a hydrophilic polar group. Many different hydrophilic groups are found in lipids which results in a very large range of potential surfactants. These include neutral functional groups such as esters, alcohols, ethers, and carbohydrates as well as ionic groups such as amines, amino acids, phosphates, sulfates, and carboxylic acids.

Previous workers[36-40] have employed several biosurfactants to release bitumen from tar sand of ambient temperatures. It was observed that during the "cold water" separation process, these biosurfactants can significantly improve the yield of bitumen in a pilot-plant. Biosurfactants can also be employed for in-situ oil recovery either by innoculating the petroleum reservoir or by producing the compound on the surface and injecting it into the well. If the surfactant is to be produced in place, it is necessary to select the microorganisms which produce good yields of biosurfactants. Improvements in yields and fermentations which use cheaper substrates than carbohydrates are needed.

SURFACTANT FORMULATIONS FOR EOR

Surfactants can be employed in different forms in enhanced oil recovery processes. The emulsions and foams have been also used in combination with other oil recovery processes. Surfactant flooding for improving oil recovery is described as follows:

<u>Macroemulsion and Microemulsion Flooding</u>: If a suitable surfactant is injected into the reservoir, it can form macroemulsions and/or microemulsions with the reservoir oil depending on the composition and reservoir conditions. Several articles have been published on the recovery of oil by microemulsion and macroemulsion flooding processes.[41-45] Among various factors, the most important factor of surfactant flooding in the form of an emulsion is the lowering of the interfacial tension (IFT) at the oil/water interface. Microemulsions are more effective in oil displacement as compared to macroemulsions because microemulsions can provide low IFT systems.

Figure 4 schematically illustrates a two dimensional view of the surfactant-polymer flooding process. After injecting a surfactant slug into the reservoir, a polymer slug is injected for mobility control. During the process, the displaced oil droplets coalesce and form an oil bank (Figure 4). Once an oil bank is formed in the porous medium, it has to be propagated through the porous medium with minimum entrapment of oil at the trailing edge of the oil bank. It is necessary to maintain the ultralow interfacial tension at the oil bank/surfactant interface for minimizing the entrapment of the oil in reservoir pores. As the injected slug moves in the reservoir, the leading edge of the oil bank coalesces with additional oil ganglia. In addition, besides interfacial tension and interfacial viscosity, another interfacial parameter which influences the oil recovery is the surface charge at the oil/brine and rock/brine interfaces. It has been shown that a high surface charge density leads to a lower interfacial viscosity, lower interfacial tension and higher oil recovery.

In EOR processes, the formation of an oil bank is very important for efficient oil displacement in porous media. This was established from studies on the injection of an artificial oil bank followed by the selected surfactant formulation which could produce ultralow IFT with the injected oil. It has been observed that the oil recovery increased significantly with the injection of an oil bank as compared to the same studies performed in the absence of an injected oil bank.[43] The oil bank formation and its propagation in a porous medium is schematically presented in Figure 5. If an early oil bank formation occurs in a porous medium, it moves through the porous medium accumulating additional oil ganglia resulting in excellent oil recovery. If late oil bank formation occurs, it will result in poor oil recovery.

In summary, various phenomena occurring at an optimal salinity in relation to enhanced oil recovery by macroemulsion and microemulsion flooding are schematically shown in Figure 6. It has been demonstrated that a maximum in oil recovery correlates well with several equilibrium and transient properties of surfactant flooding in the form of macroemulsion and microemulsion systems. Results have shown that a maximum in oil recovery, a minimum in surfactant adsorption, a minimum in apparent viscosity of the emulsion, a minimum in phase separation time, an equal solubilization of oil/brine phases in middle phase microemulsion, and a minimum in interfacial tension occur at an optimal salinity of the system.

Foam Flooding Processes: If the injection of the surfactant solution is followed by gas injection, it can form in-situ foam, which can improve oil recovery. Several aspects of foam flooding such as mechanism of foam flow in a porous medium, microscopic behavior of foam, bubble size, CO_2 foam, steam foam and oil recovery have been discussed in the literature.[45-50] Several aspects of the foam flooding process are schematically presented in Figure 7.

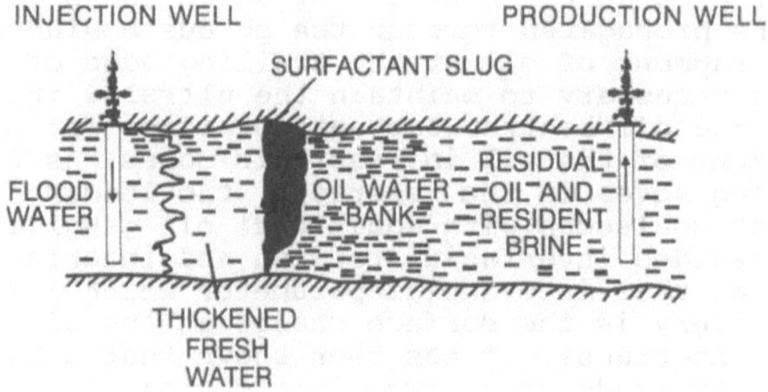

Figure 4. A Schematic Presentation of a Two Dimensional View of the Surfactant-Polymer Flooding Process.

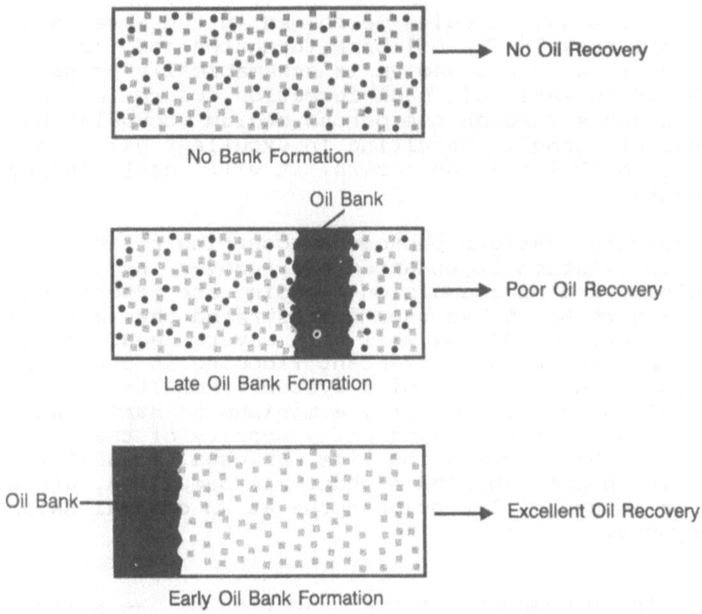

Figure 5. Formation of an Oil Bank in Relation to Petroleum Recovery.

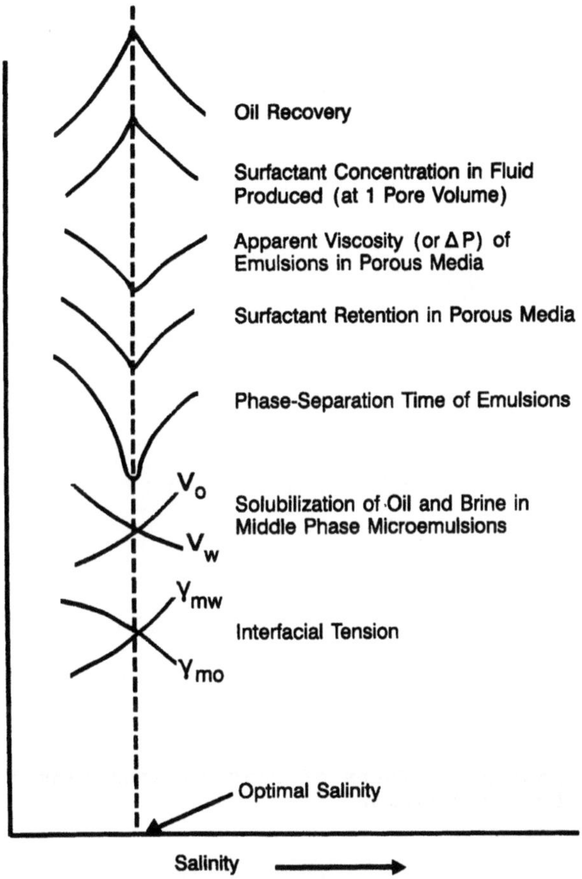

Figure 6. A Schematic Illustration of Various Phenomena Occurring at an Optimal Salinity in Relation to EOR by Surfactant Floofing.

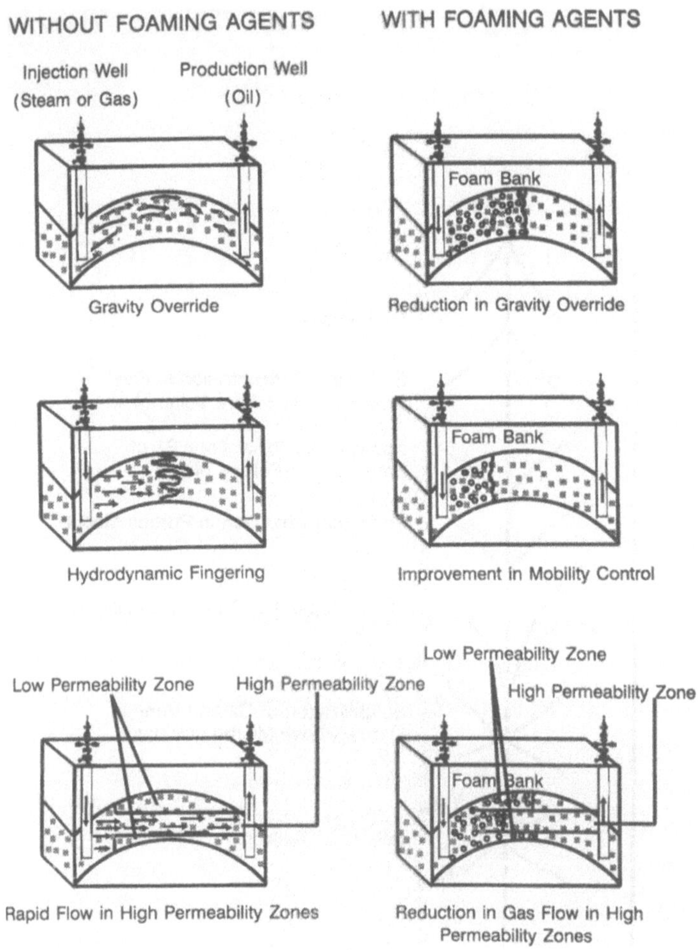

Figure 7. Possible Effects of Foam on Gas/Steam Flow in a Porous Medium.

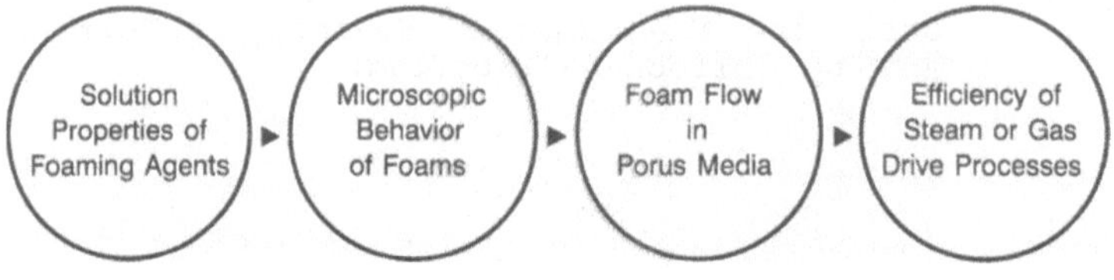

Figure 8. Interrelationship of Surface Properties of Foaming Agents with Microscopic Behavior of Foams and the Efficiency of Gas/Steam Flooding Process.

The formation of a foam bank between the displacing gas phase and the displaced aqueous surfactant solution has various possible effects on flow through porous media. The foam can act as a gravity-override reducing agent in gas/steam driven processes. As a result of in-situ foam generation in a porous media, the gravity override of injected gas/steam phase can be reduced several-folds. It can be employed as a mobility control agent. The mobility of an injected gas/steam can be controlled by the generation of in-situ foam. Foam can also be used as a selective permeability-reducing agent (e.g. blocking agent) because it is effecitive in reducing the gas permeability of porous media in high-permeability zones.

Figure 8 illustrates the interrelationship among the surface properties of foaming agents, microscopic characteristics of foams, foam behavior in porous media and efficiency of steam/gas drive processes for oil recovery. To prove the hypothesis (Figure 8), various pure surfactant systems were extensively studied knowing that these pure surfactants would not be used in the petroleum reservoir for oil displacement. Figure 9 shows a correlation of molecular properties (e.g. chain length compatibility) with surface properties of foaming solution, microscopic characteristics of foam, and heavy oil displacement by air foam and steam foam flooding processes. An air injection at the stage of surfactant solution breakthrough recovered about 2% additional oil. Replacing air by steam injection recovered nearly 12% additional oil. The steam injection presumably generates more foam in porous media as well as decreasing viscosity of the heavy oil as compared to air injection, resulting in an increase in heavy oil recovery.

For these mixed surfactant systems, a maximum in surface viscosity, a minimum in surface tension, a minimum in bubble size, and a maximum in heavy oil recovery by air foam and steam foam flooding were observed when both components of the foaming system had the same chain length[50]. The chain length compatibility effects can be explained based upon the concept of thermal motion of hydrocarbon chains in the foaming systems at an air/water interface. The mixed surfactant molecules of equal chain length form a tightly packed mixed monolayer at the air/liquid interface. When the chain lengths of the mixed surfactants are not equal, the portion of the hydrocarbon chain above the height of the adjacent hydrocarbon chain will exhibit thermal motion such as rotational, vibrational and oscillational. If these thermal motions are limited to the portion above the height of the adjacent hydrocarbon chain of the foaming agent, it would not change the packing of the molecules or the average area per molecule. The thermal motions, however, extend along the hydrocarbon chain towards the polar part of the surfactant molecule which, in turn, result in a greater area per molecule in mixed monolayers at the air/liquid interface. Moreover, as the difference in hydrocarbon chains in foaming agents increases, the expansion of the mixed monolayer also increases because the length of the segment undergoing

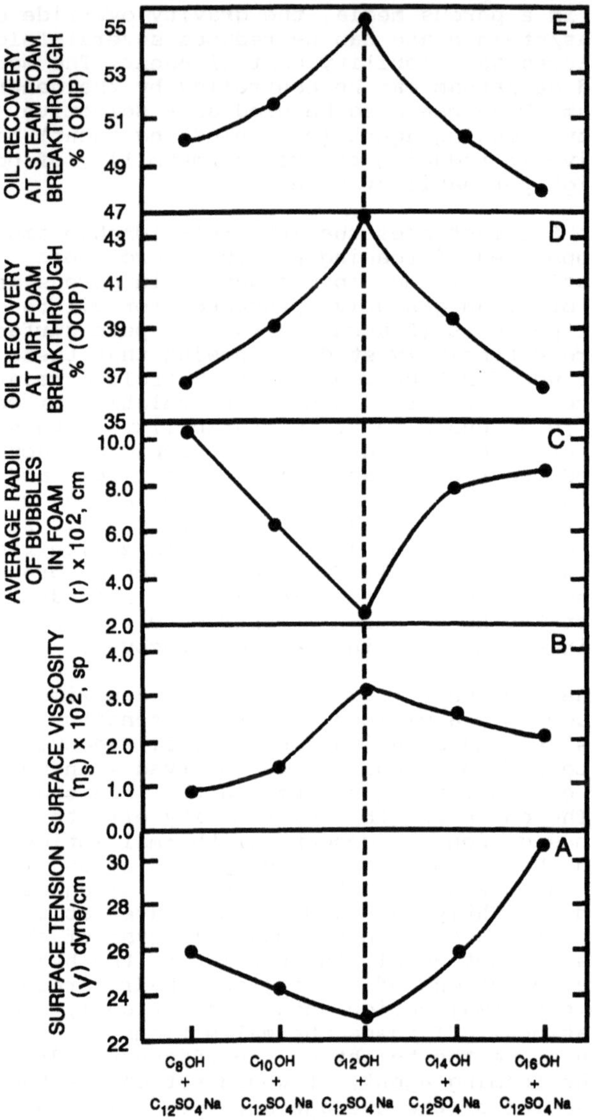

Figure 9. Effect of Molecular Properties (e.g. Chain Length Compatibility) of Mixed Surfactants on Surface Properties of Foaming Agents, Bubble Size and Heavy Oil Recovery in Porous Media.

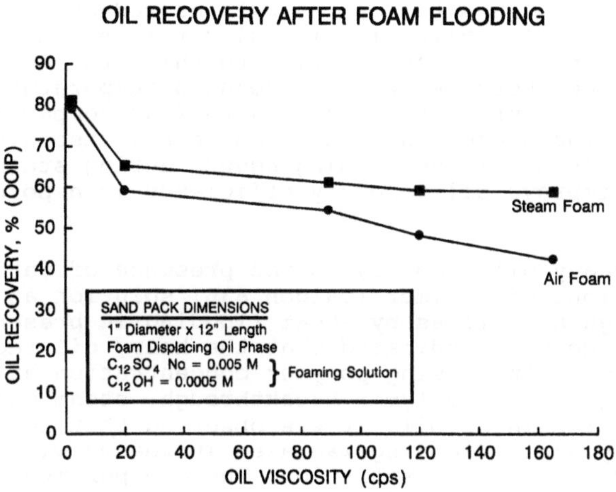

Figure 10. Effect of Oil Viscosity on Oil Displacement Efficiency by Foam Flooding.

rotation, vibration and oscillation increases. These results suggest that the molecular packing at an air/liquid interface and the surface properties of foaming solutions exhibit a striking correlation with bubble size and oil displacement in a porous medium by foam fooding.

The effect of oil viscosity on the displacement of oil is presented in Figure 10. In order to determine the oil displacement efficiency by foam flooding, the injection of gas phase was started at surfactant solution breakthrough. Both air and steam were employed to generate in-situ foams. The steam foam recovered more oil as compared to air foams. A sharp decrease in oil recovery was observed up to 25 cps oil viscosity. Beyond this, the oil recovery decreased gradually in air foam flooding as well as steam foam flooding processes.

As mentioned above, an injection of the steam in porous media generates in-situ foam as well as reduces the viscosity of oil, while oil viscosity remains the same during air injection. An addition to this, the ability to generate foam increases with increasing temperature. The steam injection produces more in-situ foam compared to air injection. Therefore, the reduction in oil viscosity and higher foaminess of the foaming agents during steam injection improve oil recovery efficiency in a porous medium.

The heavy oil recovery in the presence of various concentrations of polymer (Calgon 835) solution at polymer breakthrough as well as by steam flooding is presented in Table II. Results indicated that the heavy oil recovery increased with increasing polymer concentration at polymer breakthrough. After polymer breakthrough, steam was injected to recover remaining oil. It was observed that heavy oil recovery by steam flooding was also increased with increasing polymer concentration. These experiments were repeated by injecting 0.5 wt% Stepanflo 30 surfactant with different concentrations of polymer (Calgon 835). Results obtained are recorded in Table III. It was observed that heavy oil recovery at polymer breakthrough as well as by steam flooding (e.g steam foam flooding) was higher in the presence of surfactant as compared to that in the absence of surfactant. These results show that the surface properties play an important role in displacing heavy oil as compared to bulk properties of the injected fluids.

Based on these studies, one can select a surfactant formulation for an efficient oil recovery. The criteria for selecting a suitable surfactant for a successful oil recovery should include the parameters as described in the following sections.

SELECTION CRITERIA FOR EOR SURFACTANTS

The reservoir properties vary significantly from reservoir to reservoir or even within the same reservoir.

Table II. Heavy Oil Recovery by Steam Flooding in the Presence of Different Concentration of Calgon 835 Polymer.

POLYMER CONCENTRATION (ppm)	INITIAL OIL SATURATION (% PV)	OIL RECOVERY	
		POLYMER SOLUTION BREAKTHROUGH (% OOIP)	BY STEAM FLOODING (% OIP)
0.0	95.8	30.8	46.3
250.0	93.0	38.7	49.1
500.0	97.9	41.5	51.0
750.0	92.8	45.0	54.0
1000.0	96.9	54.7	58.1

Table III. Heavy Oil Recovery by Steam Foam Flooding in the Presence of Different Concentration of Calgon 835 Polymer.

POLYMER CONCENTRATION (ppm)	INITIAL OIL SATURATION (% PV)	OIL RECOVERY	
		POLYMER SOLUTION BREAKTHROUGH (% OOIP)	BY STEAM FOAM FLOODING (% OIP)
0.0	94.0	35.2	72.0
250.0	94.0	37.1	74.6
500.0	98.0	45.8	75.5

The rock composition, pore geometry and reservoir contents play an important role in oil displacement. Therefore, a complete study of the reservoir is required before selecting a suitable surfactant for EOR. It is recommended that the following reservoir parameters should be examined:

1. Composition of the reservoir rocks.

2. Pore geometry of the rocks.

3. Composition and type of crude such as aromatic, paraffinic, naphthenic or mixed type crude.

4. Reservoir pH and salinity.

5. Reservoir temperature and pressure.

6. Wettability of the reservoir.

7. Fluid flow behavior in petroleum reservoir

The importance of the above mentioned parameters depends on the type of surfactant formulations such as emulsions, microemulsions, foams or foams with steam etc. It is a difficult task to generalize the significance of an individual parameter. In general, the following properties are suggested to consider before selecting a surfactant for EOR processes.

<u>Solubility</u>: The solubility of the surfactants in reservoir fluid should be examined at the reservoir conditions. It is known that the solubility of the surfactants decreases with increasing salinity. For a successful recovery process, surfactants should remain soluble either in a crude oil phase or an aqueous phase, preferably at the oil-aqueous interphase under the given reservoir conditions in order to minimize its adsorption at reservoir rocks.

The solubility of ionic surfactants increases with rising temperature, whereas the solubility of nonionic surfactants decreases. If the nonionic surfactant contains an ethylene oxide chain, its solubility increases with an increasing ethylene oxide (EO) chain length in the molecule. In general, five or more EO groups in the hydrophobic molecule provide good water solubility.

The solubility of both ionic and nonionic surfactants decreases, generally, with increasing electrolyte concentration or salinity. The solubility behavior for ionic surfactants in the presence of electrolytes is more complicated as compared to that in the absence of electrolytes. The ionic surfactants are more sensitive to salinity as compared to nonionic surfactants. The selected surfactants for EOR formulation must have partial solubility in reservoir oil in order to minimize interfacial tension by partitioning at the oil/water interface, which is desired to displace oil in the reservoir.

Adsorption: During the surfactant flooding processes, the injected surfactants come in contact with reservoir contents such as crude oil, reservoir brine, rocks etc., which contain inorganic species like calcium, magnesium and alumina. As the surfactant slug travels in the reservoir, the interaction between inorganic salts and surfactants results in a loss of surfactant due to either precipitation of the surfactant or its adsorption at solid/solid interfaces. Therefore, only a part of the injected surfactant is available for oil displacement. For a successful flooding for EOR, the loss of the injected surfactant should be minimum.

The surfactant adsorption decreases with increasing molecular mass of the surfactant because the total area available to surfactant adsorption decreases. For nonionic surfactants, the adsorption occurs mostly in the form of a unimolecular layer. The adsorption decreases with increasing ethylene oxide (EO) in the surfactant molecule, but increases with the increasing length of the hydrocarbon chain. The surfactants with aliphatic hydrocarbon chains are more strongly adsorbed than surfactants containing aromatic hydrocarbon chains. For ionic surfactants, the adsorption occurs mostly in the form of multilayers. The cationic surfactants can adsorb in about 250 layers. In general, the surfactant adsorption on reservoir clay is maximum for cationic and minimum for anionic surfactants (e.g. cationic > nonionic > anionic surfactants).

The cationic surfactants strongly adsorb on reservoir silicates or quartz, which can be lowered by incorporating nonionic surfactants in the formulation. The adsorption of cationic surfactants on quartz is usually about 0.5×10^{-4} mg/cm^2. On the other hand, the nonionic surfactants adsorb slightly on quartz surfaces, but adsorption increases with raising temperature. The nonionic surfactants adsorb more on oil-wetted than water-wetted reservoir silicate surfaces.

Temperature Stability: The long-term thermal stability of injected surfactants in the reservoir environment is a must for an effective surfactant formulation. The thermal stability of several ionic and nonionic surfactants has been reported in terms of the time required for half of the surfactant to decompose (e.g. half-life).[51-53] Most of the surfactants investigated decomposed by first-order kinetics. The interfacial activity of surfactant solutions after six months at 80°C was reported by Akstinat.[54] The results showed that the stability of the surfactants was in the order: ether sulfonates > ether carboximethylates > ether sulfates > ether phosphates.

Extended research of Akstinat leads to the special findings on molecular structure suitable for EOR processes. Results reported indicate that the polyether sulfates, polycarboximethylates and polyether sulfonates of the structure type $R\ (X)_n\ Y^-\ Me^+$ are, in general, particularly suitable for oil recovery processes where R is the hydrophobic chain of a fatty alcohol, fatty acid, fatty amines or naphthanic acid; X is a polyether group-ethyleneoxide or polyethyleneoxide; Y is the polar hydrophilic head group carboxylate, sulfate, sulfonate, phosphate, etc. and Me^+ is the counter ions such as an alkali, earth alkali, amines, etc. These mixed anionic and nonionic surfactants exhibit a relatively good thermal stability as compared to anionic or nonionic surfactants alone.

In summary, a suitable surfactant should possess the following properties for a successful surfactant flood to improve oil displacement in a given petroleum reservoir.

1. Long-term stability usually 1-2 years under given reservoir conditions.

2. Partial water and crude oil solubility.

3. Ability to form an oriented monolayer.

4. Ability to stabilize the oil/water to form an emulsion.

5. Suitable to lower interfacial tension at the oil/brine interface permanently (e.g. less than $1 mN.m^{-1}$) at low surfactant concentration.

6. Enrichment at an oil/water interface under reservoir conditions.

7. Tendency to form aggregates.

8. Tendency to remain water soluble in high saline formation water.

9. Low adsorption on reservoir rocks.

10. No cosurfactant requirement to form a stable surfactant formulation.

11. No requirement of short chain alcohols to formulate stable surfactant flooding for EOR.

12. Cost effective.

In general, the surfactants selected based on the above outlined properties have reasonably good chances of success in the EOR processes.

REFERENCES

1. Gogarty, W.B. and Olson, R.W., Use of Microemulsion in iscible-Type Oil Recovery Procedure., U. S. Patent No. 3,254,714 (1962).

2. Hester, C.T., Walker, J.W. and Sawyer, G.H., Oil Recovery by Imbibition Water Flooding in the Austin and Buda Formations., J. Pet. Technol., 17(8): 919-925 (1965).

3. Higgins, R.V. and Leighton, A.J., Water Flood Performance in Stratified Reservoirs., USBM< RI-5618 (1960).

4. Sandiford, B.B., Laboratory and Field Studies of Waterfloods Using Polymer Solutions to Increase Oil Recovery., J. Pet. Technol., 16(8): 917-922 (1969).

5. Whiteley, R.C. and Ware, J.W., Low-Tension Waterflow Pilot at the Salem Unit, Marion County, Illinois, 1. Field Implementation and Results., J. Pet. Technol., 29(8): 925-932 (1977).

6. Widmyer, R.H., Frazier, G.D., Strange, L.K. and Talash, A.W., Low-Tension Waterflood at Salem Unit, Post Pilot Evaluation., J. Pet. Technol., 31(9): 1185-1190 (1979).

7. Trube, A.S., Jr., Oil Production by Primary Artificial Frontal Water Drives in the New Hope Field, Franklin County, Texas. In: Proceedings of Seventh Oil Recovery Conference, Texas Petroleum Research Committee, pp.57-75 (1954).

8. Warren, J.E. and Cosgrove, J.J., Prediction of Waterflood Behavior in a Stratifield System., Soc. Pet. Eng. J., 4(6): 149-157 (1964).

9. Hurst, W., Determination of Performance Curves in Five-Spot Waterflood., Pet. Eng., 25:B-40 (1953).

10. Hovanessian, S.A., Waterflood Calculationa for Multiple Sets of Producing Wells., J. Pet. Technol., 12(8): 65-68 (1960).

11. Gogarty, W.B., Status of Surfactant or Micellar Methods., J. Pet. Technol., 28(1): 93-102 (1976).

12. Gogarty, W.B., Mobility Control with Polymer Solutions., Soc. Pet. Eng. J., 7(6): 161-173 (1967).

13. Strange, L.K., Ignition: Key Phase in Combustion Recovery., Pet. Eng., November, pp. 105-109; December, pp. 97-106 (1964).

14. Chu, C., State of the Art Review of Steam Flood Field Projects., J. Pet. Technol., 37(11): 1887-1902 (1985).

15. Byars, C., Firefloods Blossom in North Louisiana., Oil Gas J., 68(31): 58-62 (1970).

16. Bernard, G.G., Holm, L.W. and Harvey, C.P., Use of Surfactant to Reduce CO_2 Mobility in Oil Displacement., Soc. Pet. Eng. J., 20(4): 281-292 (1980).

17. Chiang, M., Role of Surfactant Mass Transfer and the Formation of an Oil Bank in Displacement of Oil Through Porous Media., Ph.D. Dissertation, University of Florida, Gainesville, Fla. (1979).

18. Dauben, D.L. and Froning, H.R., Development and Evaluation of Micellar Solutions to Improve Water Injectivity., J. Pet. Technol., 23(5): 614-620 (1971).

19. Hill, H.R., Reisberg, J. and Stegemeier, G.L., Aqueous Surfactant System for Oil Recovery., J. Pet. Technol., 25(2): 186-194 (1973).

20. Sharma, M.K. and Shah, D.O., Use of Surfactants in Oil Recovery., In: Enhanced Oil Recovery, II Processes and Operations; Editors- E.C. Donaldson, G.V. Chilingarian and T.F. Yen, Elsevier Press, pp. 255-315 (1989).

21. Ling, T.F., Lee, H.K. and Shah, D.O., Surfactants in Enhanced Oil Recovery., In: Industrial Applications of Surfactants; Editor- D.R. Karsa, Special Publication No. 59, pp. 126-178 (1986).

22. Cayias, J.L., Hayes, M.E., Schechter, R.S. and Wade, W.H., Surfactant Aging: A Possible Detriment to Tertiary Oil Recovery., J. Pet. Technol., 28(9): 985-988 (1976).

23. Cayias, J.L., Schechter, R.S. and Wade, W.H., The Utilization of Petroleum Sulfonates for Producing Low Interfacial Tension Between Hydrocarbon and Water., J. Colloid Interface Sci., 59(1): 31-38 (1977).

24. Desai, N.N., Surfactant-Polymer Interactions in Oil Recovery Systems., Ph.D. Dissertation, University of Florida, Gainesville, Fla., pp. 70 (1983).

25. Sharma, M.K. and Shah, D.O., Interfacial Phenonena in Foam Flooding Process for Heavy Oil Recovery., 18th Intersociety Energy Conversion Engineering Conference, pp. 527-534 (1983).

26. Gale, W.W. and Sandvik, E.J., Tertiary Surfactant Flooding: Petroleum Sulfonate Composition-Efficiency Studies., Soc. Pet. Eng. J., 13(4): 191-199 (1973).

27. Kalfoglou, G., Method of Using a Combination Lignosulfates/Propyl Alkylene Glycol System as a Sacrificial Agent for Surfactant Flooding., U.S. Patent No. 5,009,268 (1990).

28. Chiwetelu, C., Hornof, V. and Neale, G.H., Enhanced Oil Recovery Using Lignosulfonate/Petroleum Sulfonate Mixtures., Trans IChemE., 60: 177-182 (1982).

29. Chiwetelu, C., Neale, G.H., and Hornof, V., Improving the Oil Recovery Efficacy of Lignosulfonate Solutions., J. Canadian Technol., 19(3): 91-99 (1980).

30. Hornof, V., Neale, G., Margeson, J. and Chiwetelu, C., Lignosulfonate-Based Mixed Surfactants for Low Interfacial Tension., Cellulose Chem. Technol., 18: 297-303 (1984).

31. Bansal, B., Neale, G. and Hornof, G., Enhanced Oil Recovery Using Lignosulfonates., Canadian J. Chem. Eng., 57: 203-210 (1979).

32. Felber, J.B., and Dauben, D.L., Laboratory Development of Lignosulfonate Gels for Sweep Improvement., Soc. Pet. Eng. J., 391-397 (December, 1977).

33. Abe, M., Schechter, R.S., Selliah, R.D., Sheikh, B. and Wade, W.H., Phase Behavior of Branched Tail Ethoxylated Carboxylate Surfactant/Electrolyte/Alkane Systems., J. Dispersion Sci. Technol., 8(2): 157-172 (1987).

34. Chase, L.K., Bryant, R.S., Bertus, K.M. and Stepp, A.K., Investigations of Mechanisms of Microbial Enhanced Oil Recovery by their Metabolic Products., National Inst. Pet. Energy Res. Bartlesville, OK., NIPER-483: Order No. DE91002210, pp. 27 (1990).

35. Updegraff, D.M., Early Research on Microbial Enhanced Oil Recovery., Dev. Ind. Microbiol., 31: 135-142 (1990).

36. Donaldson, E.C. and Clark, J.B., Proceedings on Microbial Enhanced Oil Recovery., Disrtibution Category UC-92A, CONF0205140, Published by US-DOE (1983).

37. Cooper, D.G., Zagic, J.E. and Wood, J.m., The Emulsifying and the Emulsifying Properties of Some Microbial Polysaccharides, Acta Bio., 8: 21-25 (1980).

38. Cooper, D.G., Zagic, J.E. and Gerson, D.F., Production of Surface Active Lipids by Corynebacterium Lepus., Appl. Environ. Microbiol., 37: 4-10 (1979).

39. Cooper, D.G. and Zagic, J.E., Surface Active Compounds from Microorganisms., Adv. Appl. Microbiol., 26: 229-253 (1980).

40. Moses, V., Downhole Microbially Enhanced Oil Recovery Systems., Dev. Ind. Microbiol., 31: 143-148 (1990).

41. Shaikhov, D.Z., Peshkin, O.V., Pavlyuchenko, V.I., Zlotskii, S.S. and Rakhmankulov, D.L., Microemulsion for Displacing Oil., U.S.S.R., SU 1,609,980 (1990).

42. Anghel, D.F., Phase Behavior and Interfacial Phenomena in Microemulsions Used for Enhanced Oil Recovery., Prog. Colloid Polym. Sci., 83: 188-195 (1990).

43. Sharma M.K. and Shah, D.O., Macro- and Microemulsions in Enhanced Oil Recovery., ACS Symposium Series No. 272: 149-172 (1985).

44. Chan, K.S. and Shah, D.O., The Molecular Mechanism for Achieving Ultra-low Interfacial Tension Minimum in a Petroleum Sulfonate/Oil/Brine System., J. Dispersion Sci. Technol., 1(1): 55-95 (1980).

45. Bansal, V.K. and Shah, D.O., The Effect of Ethoxylated Sulfonates on Salt Tolerance and Optimal Salinity of Surfactant Formulations for Tertiary Oil Recovery., Soc. Pet. Eng. J., 18(3): 167-172 (1978).

46. Danzik, M., Viscosity Control Additives for Foaming Mixtures in Petroleum Recovery., U.s. Patent No. 5,000,262 (1991).

47. Podustov, M.A., Goncharova, Z.V., Pravdin, V.G. and Tosunov, E.M., New Foaming Compositions for Treatment of the Critical Zone of Oil Wells., Neft. Khoz., 9: 49-52 (1990).

48. Pakki, V.I. and Goncharov, V.N., Breakdown of Stable Foams in the Recovery, Processing, and Transport of Gas., Neft. Gazov. Prom-st., 3: 40 (1990).

49. Sharma, M.K. and Shah, D.O., Surface Properties of Foaming Agents in Relation to Foam Flow in Porous Media., 184th Natl. Meet., ACS, Washington, D.C., August 21-26, GEOC 039 (1982).

50. Sharma, M.K. and Shah, D.O., Effect of Oil Viscosity on Recovery Processes in Relation to Foam Flooding., J. Am. Oil Chem. Soc., 61(3): 585-590 (1984).

51. Sharma, M.K., Shah, D.O. and Brigham, W.E., The Influence of Temperature on Surface and Microscopic Properties of Foaming Solutions in Relation to Fluid Displacement Efficiency in Porous Media., AIChE J., 31(2): 222-228 (1984).

52. Handy, L.L., Behavior of Surfactants in EOR Applications at High Temperatures. In: Proceedings, European Symposium on EOR, Bournemouth, England, Sept. 21-23, Elsevier, Lausanne, pp. 149-160, (1981).

53. Handy, L.L., Amaefule, J.O., Ziegler, V.M. and Ershaghi, I., Thermal Stability of Surfactants for Reservoir Applications., Int. Symp. on Oil Field and Geothermal Chemistry, Houston, Texas, January 22-24, pp. 61-70, SPE 7867 (1979).

54. Akstinat, M.H., Surfactants for EOR Processes in High Salinity Systems- Production, Selection and Evaluation, In: Proceedings, European Symposium on EOR, Bournemouth, England, Sept. 21-23, Elsevier, Lausanne (1981).

COSURFACTANT-ENHANCED ALKALINE/POLYMER FLOODS FOR IMPROVING RECOVERY IN A FRACTURED SANDSTONE RESERVOIR

M. R. Islam

Petroleum Development Laboratory
University of Alaska Fairbanks
Fairbanks, Alaska

This paper presents a new recovery technique for improving waterflood performance in a fractured sandstone reservoir. The presence of fractures leads to very early water breakthrough and makes it difficult to drain the matrix of the fractured reservoir. Consequently, these fractured reservoirs are left with considerable amounts of unproduced oil. An effective technique for plugging the fractures and simultaneous treatment of the matrix to produce oil will lead to great improvement in oil recovery from fractured reservoirs. One such technique is presented in this paper.

A new technique using CO_2-activated plugs of sodium orthosilicate is used to plug fractures in sandstone cores. This is followed by injection of chemical slugs, such as alkali/surfactant and polymer/alkali/surfactant. A series of experimental runs were conducted on artificially fractured Berea sandstone cores. It is shown that recovery as high as 75% of the oil in place is possible with the novel treatment as compared to less than 10% recovery with waterfloods. Results are presented for various slug sizes.

INTRODUCTION

Fracturing is a very common phenomena in sandstone and limestone reservoirs. Such fracturing occurs due to uneven lift and subsidence across an oil reservoir over geologic time. Of course, the intensity of fracturing depends on the brittleness of the rock itself. Most of the natural fractures are open or only partially plugged by natural mineral precipitations. The presence of fractures affect the oil production mode intensely. Even though open fractures contribute little to the total pore volume of the reservoir, they act as very high conductivity channels in the reservoir. While most of the oil in place is situated in the matrix of a fractured reservoirs, this matrix is often found to be very tight and, in most cases, shows very low permeability. Due to this large contrast in matrix and

Particle Technology and Surface Phenomena in Minerals and Petroleum
Edited by M.K. Sharma and G.D. Sharma, Plenum Press, New York, 1991

reservoir permeability, it is very difficult to produce a fractured reservoir with viscous displacement by water or gas. Consequently, gravity drainage, solution gas drive, and water imbibition are the most common recovery mechanisms prevalent in fractured reservoirs. However, in many reservoirs, these recovery mechanisms are almost non-existent. Therefore historically waterflooding has been conducted to recover oil from fractured sandstone reservoirs. In order to reduce conductivity of the fractures, a combination of cationic and anionic polyacrylamides with possible aluminum citrate cross-linking is used [1,2]. Cationic polyacrylamide is injected first and enters the fractures due to high conductivity of the fractures. This is followed by the addition of anionic polyacrylamide which attaches to the cationic polymer. Often aluminum citrate is added to crosslink the polymer molecules, forming a colloidal dispersion gel which creates resistance to flow in the fractures. This is typically followed by water injection. In order to ensure water imbibition, a wettability adjustment agent consisting of a blend of anionic polymer and sodium tripolyphosphate [3,4,5] is often injected prior to waterflooding.

In this paper, a new technique for plugging fractures is also presented. Also, surfactant-enhanced alkali/polymer flood is alternated with water injection in order to optimize oil production. This technique uses a shorter sequence of chemical injection and therefore increases the possibility of plugging the fractures.

EXPERIMENTAL SET-UP

All experiments were carried out by using Berea sandstone cores. Fractures were simulated by using two blocks of Berea sandstone as shown in Figure 1. These blocks were separated with several 1 sq. cm 0.5 mm thick plates. This ensured high permeability of the fracture as well as minimal width of the fracture (estimated to be 0.05 cm). These rectangular blocks were cast in a stainless steel core using a low melting alloy. In order to have access to the core, holes were drilled through ports B, C, D, F, G, and H. A special technique had to be used to saturate the core with crude oil. As shown in Figure 1, three injection ports were located on top of the core holder while three production ports were placed in the bottom. In order to saturate the core uniformly, the following steps were taken. A vacuum was pulled in the core holder overnight. Following this, port A was exposed to a brine (2% NaCl) solution and brine was imbibed into the core. In order to ensure complete saturation of the core, water was injected through ports B, C, and D and was produced from ports F, G, and H. An accurate material balance was used to determine pore volume of the core. Following water saturation, oil saturation was carried out through port B with production through port F. The oil injection was continued until water cut fell below 1% of the total effluent rate. This was followed by oil injection through port D while producing through port H. Similar to the previous case, oil saturation was continued until water production dropped below 1% of the total production. Finally, oil injection was carried out through port C, and production was carried out through port G. When water production dropped below 1% of the total flow through the port G, ports F, G, and H were opened. This procedure allowed one to obtain an initial oil saturation of 80%. At this point the core holder system was ready for waterflood and chemical treatments. Waterflooding or chemical injection was carried out with a constant rate, positive displacement syringe pump which displaced machine oil to push floating pistons which displaced desired fluids. The production well was connected to a separator. The volume of gas was measured with a flow meter while liquid volume was collected in a fraction collector. Chemicals used in different runs are listed in Table 1. Interfacial tension between the crude oil and different chemical solutions were obtained by using a spinning drop interfacial tensiometer.

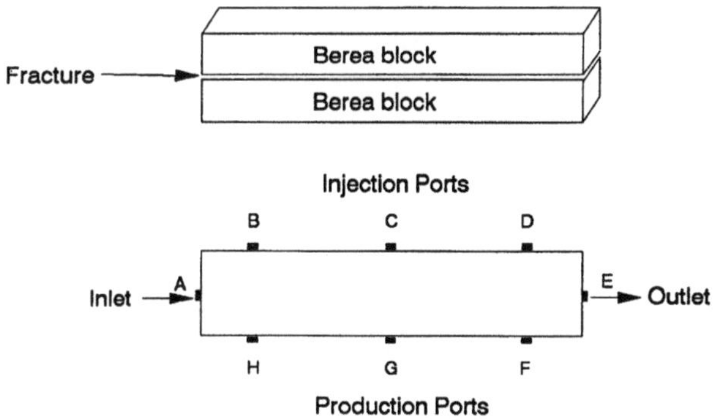

Figure 1. Schematic of the Core and Fracture System

RESULTS AND DISCUSSION

Gelation Procedure

In order to obtain the most effective gelation, the pH of the the orthosilicate solution was brought down to 9.4 by the addition of HCl. Islam[6] reported this pH to be optimum for CO_2 activation. Also, the CO_2 requirement was reported to be four times the volume of the sodium orthosilicate solution In the present study the target for creating *in situ* plugging was the fractures. Therefore, the amount of solution required for optimum gelation had to be determined through experimentation in the presence of a fracture. In order to determine optimum slug volume, four runs were conducted. In all the runs, different volumes of sodium orthosilicate solution were injected after saturating the core with water following the procedure outlined earlier. No oil saturation was conducted for this series of runs. A volume of CO_2 four times the volume of sodium orthosilicate solution was then injected for each case. All the cores were left intact for gelation overnight. Then, a waterflood at the same rate was carried out for each slug volume of sodium orthosilicate. The goal was to achieve permeability close to the original permeability of the sandstone blocks (in the absence of fractures). Results of these runs are shown in Figure 2. Note that the volume of the chemical solution is expressed in pore volume of the fracture itself. The fracture volume was approximated to be 11.5 ml. As can be seen in Figure 2, the effectiveness of gelation improved as the volume of sodium orthosilicate slug increases. However, beyond 2 pore volumes of the fracture, permeability is reduced severely. This is a source of loss of injectivity in a field situation. From these results, it was decided to use a sodium orthosilicate solution of two times the pore volume of the fractures for the rest of the coreflood tests.

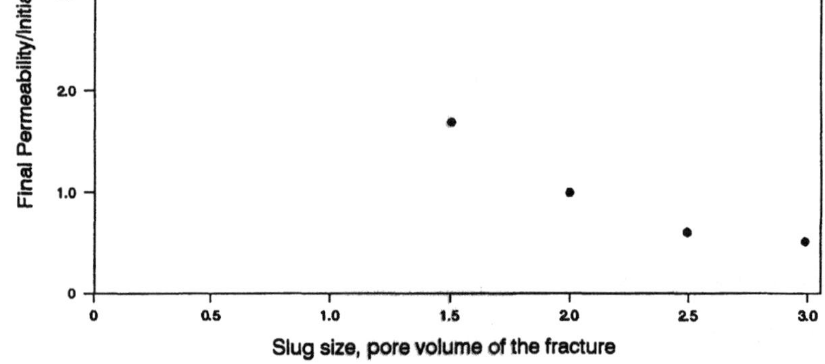

Figure 2. Effect of Sodium Orthosilicate Slug Size on Overall Permeability

Displacement of Oil in the Matrix

In order to improve recovery from the matrix, chemical slugs of alkali, polymer, and surfactant were used. Sodium carbonate was used as an alkaline phase. It is considerably inexpensive compared to NaOH and is excellent as a buffering agent. It is also considerably less reactive [7,8,9] and propagates several times faster than the NaOH solution at a pH of 11.4[10]. Taylor et al.[11] were the first researchers to report dynamic interfacial tension (IFT) induced by Neodol, a commercially available alcohol ethoxy sulphate, in the presence of alkaline solutions. This is particularly helpful because, as they demonstrated, such a dynamic IFT increases oil recovery substantially irrespective of how high the final IFT is. Later, Islam and Chakma[12] demonstrated through numerical simulation that the IFT does not have to be ultralow throughout the displacement test. In fact, a dynamic IFT helped recover more oil than a constant ultralow IFT. Similar dynamic IFT has been observed in the present work in the presence of alkali, Neodol, and polymer. The addition of polymer changed the dynamic IFT behavior only slightly. The lowest dynamic IFT was obtained at 0.05% Neodol, 1% Na_2CO_3 and 0.05% polymer. Consequently, it was decided to use these concentrations for following displacement runs.

Waterflood

Run 1 was conducted using waterflood alone. No gelation or any other chemical agent was used in this run. As expected, the water breakthrough took place very quickly. At the end of the displacement test, only 7% of the oil-in-place was recovered. For the sake of comparison, oil recoveries at 5% oil cut were used as standard values. Note that in this experimentation,dead oil was used. Consequently, this 7% recovery does not include any production through primary depletion which is likely to happen in a pressurized reservoir. Results of this run show why displacement-type recovery has not been used in naturally fractured reservoirs.

Waterflood with Silica Gel Plugging

Run 2 was conducted by injecting two fracture pore volumes of sodium orthosilicate and HCl solution at a pH of 9.4. This was followed by a CO_2 injection. The CO_2 injection rate was slow (20 ml/hr) and a total volume of 90ml was injected. During gel and CO_2 injection, the core was left intact overnight. The same procedure was used for Runs 2 through 10. For Run 2 this gelation period was followed by a waterflood. The injected water salinity was the same as the interstitial water (2% NaCl). During waterflood, injection pressure was very high. Such a high injection pressure was also reported by Islam and Farouq Ali[13], who used CO_2-activated silica gel for plugging bottom-water zones of oil reservoirs. Figure 3 shows the waterflood recovery performance (pressure and oil recovery data). As can be seen in this figure, injection pressure continued to rise until shortly before water breakthrough. Water breakthrough took place after 20% of the pore volume injection. Following water breakthrough, oil cut dropped rapidly. However, as the oil cut dropped below 5%, 29% of the oil-in-place was recovered. This indicates that close to 200% of additional oil is recovered over conventional waterflooding which gave a recovery of only 7% of the oil-in-place.

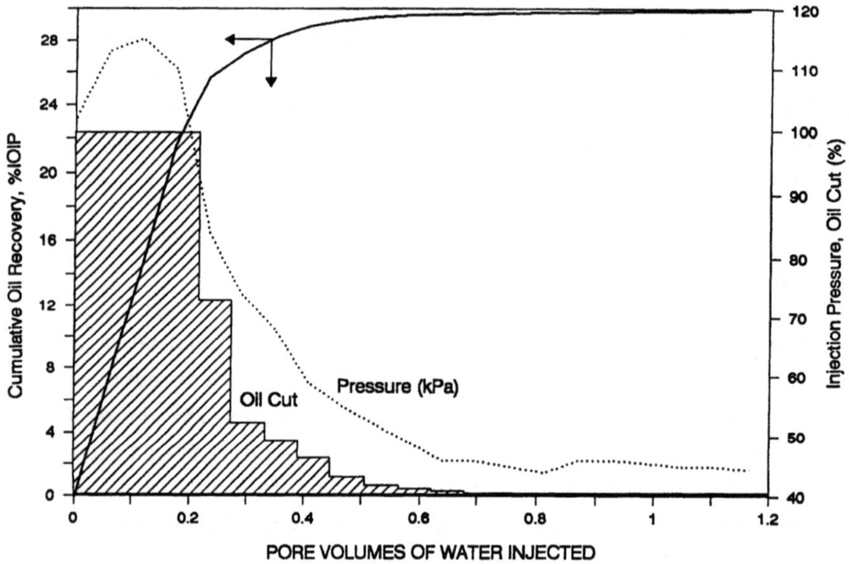

Figure 3. Recovery Performance of Waterflood Following Silica Gelation

Recovery with Alkali/Surfactant

Runs 3 through 6 were conducted using alkali/Neodol as the displacement fluid following silica gelation. For all the runs, a 1% Na_2CO_3 and 0.05% Neodol concentration was used. As mentioned earlier, the dynamic IFT was the lowest at this chemical concentration. Chemical slug sizes of 20%, 40%, 80%, and 100% of the pore volume were used for Runs 3 through 7, respectively. Recovery results are compared in Figure 4. Note that oil recovery increases substantially as the slug size is increased from 20% to 80% of the pore volume. However, beyond 80% slug size, no additional oil is recovered by increasing the slug size. As the chemical slug propagates, the oil is displaced immiscibly. However, due to highly unfavorable mobility ratio, viscous fingering takes place. This is why chemical breakthrough takes place almost at the same time as conventional water injection. However, following breakthrough as the residual oil comes in contact with chemical solutions, IFT decreases to an ultralow value and oil continues to be mobilized. That is why chemical flooding cases do not show rapid decline in oil cut after breakthrough of the injected fluid. However, if the slug size is too small (20% of the pore volume), the chase water is injected as soon as the breakthrough take place. The injected water displaces the chemical slug which finds its way though widening viscous fingers. Therefore at the inlet end no residual oil comes in contact with dynamic ultralow IFT-induced chemicals. Consequently, a 20% slug recovers relatively small amounts of additional oil. The average of chemical slug increases as the slug size is increased as more oil comes in contact with fresh chemicals even after initial breakthrough. However, for very large slug volumes oil recovery approaches that of continuous chemical injection. Increasing the slug size further does not improve recovery. Figure 5 compares oil recoveries for various slug sizes. In order to be able to compare at a given water-oil ratio, 5% oil cut was chosen as the standard point. The improvement in oil recovery for increasing slug size declines somewhere between 40% and 80% of the slug size. For economic purposes an optimum slug size would be between these two points.

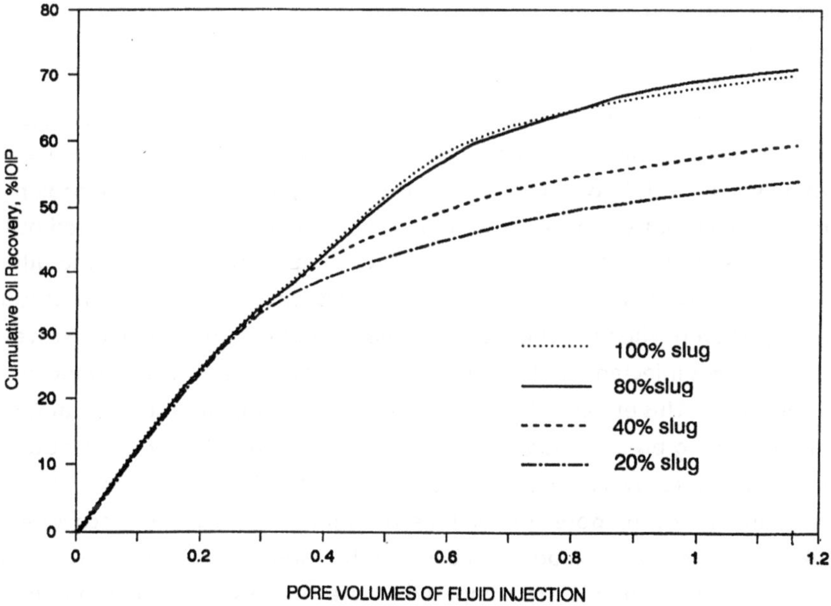

Figure 4. Comparison of Oil Recoveries for Various Slug Sizes of Alkali/Surfactant

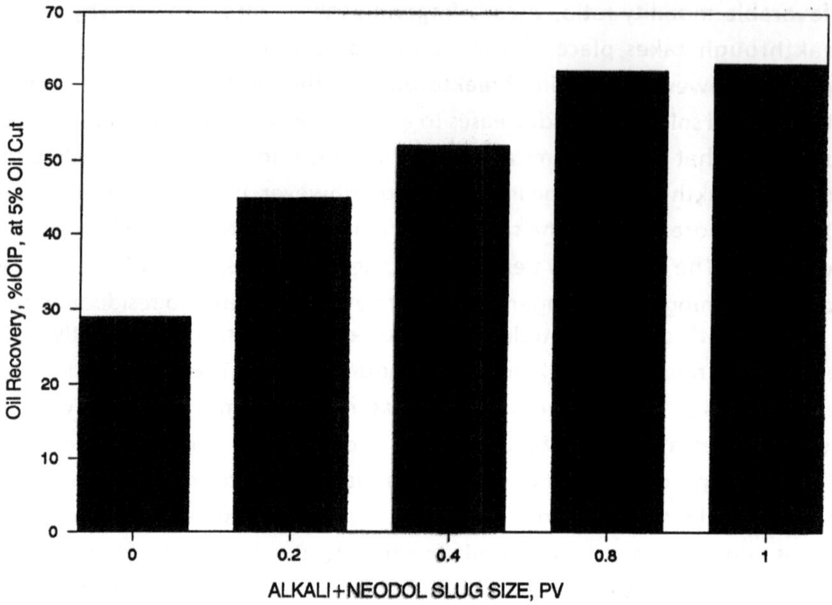

Figure 5. Comparison of Oil Recoveries (at 5% oil cut) for Various Alkali/Surfactant Slub Sizes

Recovery with Polymer, Alkali, and Neodol

Runs 7 through 10 were conducted using 0.05% Neodol, 0.05% polyacrylamide, and 1% Na_2CO_3. Slug sizes of 20%, 40%, 80%, and 100% of the pore volume were used for Runs 7 through 10, respectively. This viscosity of this solution was 65 mPa.s. Obviously, this high viscosity led to a favorable mobility ratio. Results of these runs are shown in Figure 6. Note that for all cases, a favorable chemical-oil mobility ratio led to delayed breakthrough. However, when polymer was followed with the brine solution, profuse fingering took place despite the fact that the polymer solution is miscible in water. As pointed out by Islam[6], this resulted in dilution of the chemical slug. However, even the diluted solution gives rise to favorable mobility ratio. This will explain why breakthrough was delayed even when the chemical slug was 20% of the pore volume. However, if a larger slug is injected, no dilution of the chemical takes place and smooth slug flow continues. This effect not only delays breakthrough, but also maintains high oil rate following breakthrough. However, as observed before with Neodol and alkali solutions, recovery becomes insensitive to the slug size when a slug size larger than 80% of the pore volume is used. Figure 7 compares oil recoveries for all alkali, polymer, and Neodol injections with various slug sizes. Once more, recoveries at oil cut of 5% are compared. Note that a very large improvement over conventional water injection takes place even when the chemical slug size is

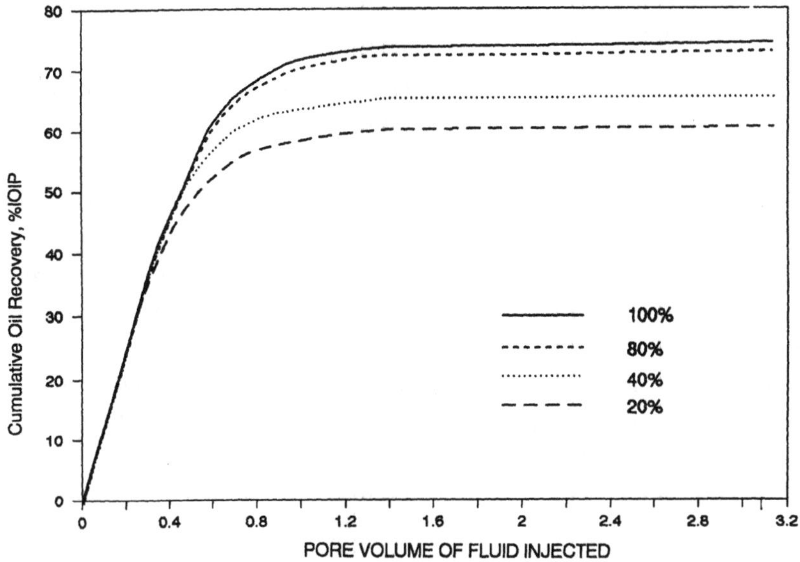

Figure 6. Comparison of Oil Recoveries for Various Slug Sizes of Alkali/Surfactant/Polymer

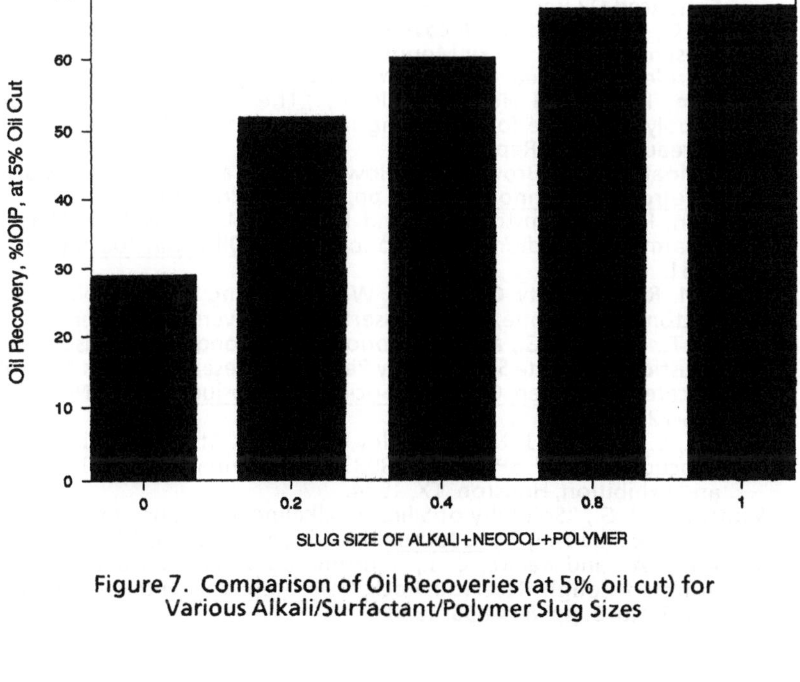

Figure 7. Comparison of Oil Recoveries (at 5% oil cut) for Various Alkali/Surfactant/Polymer Slug Sizes

relatively small (e.g., 20% of the pore volume). Also, all of these cases recover considerably higher amounts of oil than corresponding cases of alkali and Neodol injection.

SUMMARY AND CONCLUSIONS

A new recovery technique using CO_2-activated silica gel formation followed by chemical slug injection is developed for recovering medium light oils from fractured reservoirs. A mixture of alkali and surfactant (Neodol) or of alkali, surfactant, and polymer are used as chemical slugs. Initial tests indicate minimum dynamic interfacial tension for a solution of 1% Na_2CO_3 and 0.05% Neodol and 1% Na_2CO_3, 0.05% Neodol and 0.05% polyacrylamide. Chemical slugs of 40% to 80% of the pore volume was found to be optimum for oil recovery. Increasing the slug size beyond 80% did not produce additional oil. In all cases, the combination of chemical flooding and CO_2-activated silica gel formation were found to be adequate for producing in fractured reservoirs which show very poor displacement efficiency under conventional waterflooding (7% of the IOIP). Recovery was as high as 75% of the oil-in-place using the technique presented in this work.

REFERENCES

1. Hochandalel, S. M., Lunceford, M. L., and Farmer, C. W., "A Comparison of 31 Minnelusa Polymer Floods with 24 Minnelusa Waterfloods," SPE/DOE Paper 20234 presented at the SPE/DOE Seventh Symposium on Enhanced Oil Recovery, Tulsa, OK, 1990.
2. Mack, J. C., "Improved Oil Recovery - Product to Process," SPE Paper 7179 presented at the Rocky Mountain Regional Meeting of the Society of Petroleum Engineers, Cody, WY, 1978.
3. Johansen, R. T., and Heemstra, R. J., "The Effectiveness of Sodium Tripolyphosphate for Improving Injection Rates of Waterfloods," U.S. Bureau of Mines Report 6557.
4. Sloat, B., and Brown, M., "How to Flood a Tight Sand," Journal of Petroleum Technology, Oct., pp. 1119 - 1128, 1968.
5. Johansen, R. T., Heemstra, R. J., and Hembree, L., "Low-Cost Waterflood Treatment of Sodium Tripolysphosphate," Oil & Gas Journal, June 16, 1961.
6. Islam, M. R., "Mobility Control in Waterflooding Oil Reservoirs with a Bottom-Water Zone," Ph.D. dissertation, University of Alberta, 1987.
7. Lieu, V. T., Miller, S. G., and Staphonos, S. J., "Long-term Consumption of Caustic and Silicate Solutions by Petroleum Reservoir Sands," in Soluble Silicates, American Chemical Society Symposium Series, No. 194, pp. 215-226, 1982.
8. Mohnot, S. M., Bae, J. H., and Foley, W. L., "A Study of Mineral-Alkali Reactions," Part I, SPE Paper 13032, 59th Annual Technical Conference and Exhibition, Houston, TX, 1984.
9. Southwick, J. G., "Solubilty of Silica in Alkaline Solutions: Implications for Alkaline Flooding," SPE Journal, Vol. 25, No. 6, pp. 857-864, 1985.
10. Jensen, J. A., and Radke, C. J., "Chromatographic Transport of Alkaline Buffers through Reservoir Rock," SPE Reservoir Engineering Journal, Vol. 3, No. 3, pp. 849-856, 1988.

11. Taylor, K. C., Hawkins, B. F., and Islam, M. R., "Dynamic Interfacial Tension in Surfactant Enhanced Alkaline Floodings,"
12. Islam, M. R., and Chakma, A., "Mathematical Modeling of Enhanced Oil Recoveyry by Alkali Solutions in the Presence of Cosurfactant and Polymer," Journal of Petroleum Engineering Science, in press, 1990.
13. Islam, M. R., and Farouq Ali, S. M., "Improving Waterfloood in Oil Reservoirs with Bottomwater," SPE Paper 16727 presented at the SPE Annual Technical Conference and Exhibition, Dallas, TX, 1987.

FOAM FLOW BEHAVIOUR IN POROUS MEDIA IN RELATION TO ENHANCED OIL RECOVERY (EOR)

Dhirendra Kumar

Lunglei Government College
Lunglei-796701, Mizoram (India)

Several foam properties in relation to enhanced oil recovery by foam flooding are presented in order to understand the mechanism of fluid/oil displacement in porous media. The fluid displacement experiments were conducted under different conditions of gas injections and types of the porous media. It was observed that the breakthrough time and fluid displacement efficiency were increased with increasing surfactant concentration. At low surfactant concentration, the effective gas mobility was high, and decreased with increasing concentration. Foams were effective in reducing the gas mobility in the porous media. For high fluid displacement, the effective gas mobility should be minimal. As the permeability increased, the flow of foam through porous media also increased. The breakthrough time and fluid displacement efficiency were correlated with slug size. Results indicated that the breakthrough time (BT) and fluid displacement efficiency (FDE) or fluid recovery (FR) were increased with increasing slug size. Upon increasing temperature, the surface tension and bubble size decreased which, increased the fluid displacement efficiency. The effective gas mobility in the presence of a surfactant solution decreased with increasing temperature. The effect of applied pressure on breakthrough time and fluid displacement showed a gradual decrease in fluid displacement efficiency/oil recovery with increased applied pressure in the absence of foam. An increase in the fluid/oil displacement efficiency was observed at low pressures and in the presence of foam. In the presence of foam, the breakthrough time and fluid recovery were increased. Fluid/oil displacement efficiency increased with increasing oil viscosity. More viscous oils were recovered in lower amount compared to the less viscous oils in the presence of foam.

INTRODUCTION

Foam is composed of gas bubbles dispersed uniformly throughout a continuous liquid phase, and can be treated as a homogeneous fluid with both variable density and viscosity. Foam can be generated in-situ by

Particle Technology and Surface Phenomena in Minerals and Petroleum
Edited by M.K. Sharma and G.D. Sharma, Plenum Press, New York, 1991

the injection of a surfactant solution followed by steam/gas injection. In 1958, Bond and Holbrook[1] proposed that the oil recovery agent may be a mixture of gas and surface active agent solution. Foam has been investigated as an oil displacing agent[2,3]. Foam has also been used in gas reservoirs to reduce gas leaks[4,5].

Surfactants play an important role in the formation and stability of foams. Investigators[6,7] have determined foam stability by measuring the half-life (e.g. $t_{1/2}$) of the foam. Half-life is the time required to reduce foam volume to half of its initial value. It has been demonstrated that the foam stability (i.e.half-life) decreased with increasing temperature, whereas the foaminess of the surfactant solution increased with temperature. It is likely that these properties of foam depend on the molecular structure and concentration of the surfactant at the gas/liquid interface. Comparison of the results of static foam stability with that of the dynamic behavior of foam in porous media revealed that the foam stability is not required for efficient fluid displacement or a decrease in the effective air mobility in a porous medium. Moreover, the ability of the surfactants to produce in-situ foam was one of the important factors in the displacement of the fluid in a porous medium.

The bubble size influenced the efficiency of the foam flooding process. Mixed surfactants of equal chain length produced smaller foam bubbles as compared to mixed surfactants of unequal chain length[8]. It was observed that the smaller bubbles displaced more fluid as compared to the large size bubbles. The rate of increase of bubble size was found to be smaller for surfactants containing more surface-active material compared with surfactants having less surface-active material. The bubble size decreased with increasing surfactant concentration; however beyond the CMC, the bubble size remained almost constant for both pure and commercial surface active agents[9].

Foam is the only known compressible non-Newtonian fluid with both variable density and viscosity. The viscosity of a foam is greater than that of either of its components. The foam viscosity changes with the concentration of the foaming agents as well as with changes in pressure. The surfactant solution with high surface viscosity produces a stable foam film due to the tight packing of the surfactant molecules. Surface viscosity of foaming solutions has been measured by Mannheimer and Schechter[10-12]. The concentration of the surfactant can affect the surface viscosity. These observations indicate that the surface viscosity increases with increasing surfactant concentration, and beyond the CMC the surface viscosity remains almost constant.

It was observed that the low surface tension foaming solutions tend to produce stable foams at lower concentration of surfactants. The surface tension of the foaming solutions was measured by Klevens[13]. He observed that the addition of the brine to the surfactant solutions resulted in an increase in surface tension of the system. Surface tension increased with increasing difference in chain length of mixed surfactants.

Further studies showed that a decrease in surface tension was obtained with increasing temperature[14]. As the temperature was raised, the kinetic energy of the molecules increased, resulting in a decrease in attractive forces between the molecules which in turn, reduced the surface tension of the surfactant solutions. It was concluded that, as the surface tension decreased the fluid displacement efficiency in porous media increased.

The foaming ability of surface active agents depends upon the critical micelle concentration (CMC). The micelles are formed beyond a certain critical concentration of surface active agent called the "critical micelle concentration"or CMC. It has been reported by Corrin and Harkins[15] that the addition of long chain alcohols or electrolytes lowers the CMC of the ionic surfactants. In the absence of additives, the CMC of ionic surfactants generally increases with increasing temperature[16]. However, the combined effect of brine and alcohol in reducing the CMC of mixed foaming agent is greater than the effect of temperature in increasing the CMC. Thus maintaining the concentration of the mixed foaming agents above the CMC.

Foam stability has been the most studied foam property. Early reports of highly stable bubbles and foams may be found in the literature[17]. It is a well known fact that pure liquids do not form foams, nor, usually, do solutions of similar liquids or aqueous solutions of highly hydrophilic solutes[18]. Molecules of superior foaming agents contain both lyophilic and lyophobic groups and are, therefore, adsorbed at the interface when dissolved in a solvent[19]. From the experimental work of Burcik[20], it was concluded that foam stability is favored by (i) low solution surface tension relative to that of the pure solvent; (ii) moderate rate of surface tension lowering; and (iii) high bulk or surface viscosity. Further work by Miles and Ross[21] indicated that the low stability is observed with increasing pH of the solution. Changes in temperature may alter the viscosity and the surface tension of a solution and are, therefore, important to stability. When the foam height in the column becomes stabilised i.e., when the rate of collapse equals to the rate of formation, the stability is found to be a function of the foam volume and the volume rate of gas bubbling through the solution.

Foam stability in the presence of crude oil and pure hydrocarbons has been investigated as a function of chain length of olefin sulphonates and electrolyte concentration[22]. Interactions between aqueous foam films and emulsified oil droplets has been studied using transmitted light, incident light interferometric and differential interferometric microscopic techniques. Three phase foam stability has been discussed by numerous authors[23-26]. However, the mechanisms by which oil affects foam stability are still under investigation. The effect of oil upon foam stability has been explained in general terms through the mechanism of oil spreading phenomena[27]. It is suggested that during the process of lamella thinning, the oil droplets are squeezed between the film surface and spread on the film surfaces.

The dependence of foam stability upon surfactant concentration is well known. Specifically, above a certain surfactant concentration (after CMC), the stability of the foam increases sharply with surfactant concentration. At these high surfactant concentrations spheroidal micelles are formed in the bulk phase[28], which, above a certain surfactant concentration may achieve a regular periodic structure. This periodic structuring occurs due to a spatial ordering of the micelles in periodic potential minima arising from the balance between the attractive and repulsive long range micellar interactions[29]. It was concluded that firstly, after a certain surfactant concentration (above the CMC) the stability of the foam film will increase due to the additional stabilizing force of the micellar periodic structure in the film. Secondly, increasing the electrolyte concentration will decrease the repulsive forces between the micelles in the film thus inhibiting the formation of a micellar periodic structure in the film core. The effect of electrolyte will, of course, differ for anionic and nonionic surfactants. For similar micelle concentrations, the effect of electrolyte will be more severe for the anionic surfactant as it will suppress the electrostatic double-layer repulsive forces acting between the surfactant aggregates. For nonionic surfactant, the repulsive force between micelles is a steric force rather than an electrostatic force, such the electrolyte has less of an effect.

Roof[30] investigated the conditions that must be met in order that the oil emerging from a water-wet constriction seperate (Choke-off, Snap-off or pinch-off) into a droplet in a larger channel. Such flow of water and oil in a water-wet system is similar to the displacement of an aqueous surfactant solution by air in a water-wet porous medium. If snap-off occurs in the latter case, seperate air bubbles and a network of air-liquid interfaces will be produced.

The surfactant solution in the reservoir can alter the interfacial tension at solid/liquid and gas/liquid interfaces, which in turn influences the capillary number. For a capillary controlled displacement, the leading spherical interface becomes unstable if its capillary pressure is less than the capillary pressure at the throat. Roof[30] considered toric pores and computed the equilibrium location of the leading front as a function of tore radius. The front (head meniscus) is always at least seven pore radii from the throat of the constriction before snap-off (choke-off) can occur. He made an experimental verification of the theoritical analysis. His experimental observations showed that irregularity in the pore seems to be necessary for a snap-off process to occur within reasonable periods of time.

Mohanty et al.[31] made an energy stability analysis of neck and head menisci. They determined the minimum ratio of pore body radius to adjacent throat radius (rb/rt) required for foam snap-off to occur. For most shapes considered, snap-off occurs when this ratio is greater than three. In both Roof's and Mohanty's work, the criteria for snap-off were independent of interfacial tension and other fluid properties. For a snap-off of oil or injected foam to occur in a flow channel, the leading fluid interface must advance into cavities considerably larger than pore throats. The relative sizes of pores and throats, rather than their absolute sizes, thus influence choke-off.

Wardlaw and Cassan[32] commenting on the effect of pore structure recovery efficiency contend that "In regular packing of equal spheres (cubic, rhombohedral-hexagonal), no capillary trapping can be expected because a nonwetting phase entering and filling a pore will be constrained by the pore wall on the downstream side before the curvature can expand enough to cause snap-off at the upstream pore-neck.

The reports of the enhanced oil recovery projects concluded that the reservoir heterogeneities is the most frequent cause for failure of enhanced oil recovery processes which involve foam and surfactant flooding. It was observed that the reservoir heterogeneity was much more dominant than expected. Geologic and permeability heterogeneities were the most probable cause of the low recovery efficiency realized by Goodrich and Watson[33]. The wettability of the pore surface affects the recovery efficiency, but this effect is poorly understood because it is almost impossible to determine which portions of the surface are oil-wet and which ones are water-wet in the subsurfaces. The wettability of the reservoir rocks can be influenced by the injection of surfactant solution for foam flooding.

Foam process to improve oil recovery has been studied by various investigators[34-40]. In their experiments, a water soluble surface active agent with foam producing ability was injected into an under ground formation as an aqueous slug. This slug was followed by gas or steam injection to produce foam within the petroleum reservoir. A review of foam research and applications of foam in the petroleum industry was compiled by Marsden[41].

Foam is formed when an aqueous solution of foaming agent and gas are injected into a porous medium either simultaneously or intermittently. A difference of opinion exists among the researchers in regard to the behaviour of foam flow in porous media. Some investigators have considered the flow of foam as a single phase of high viscosity fluid, and others have treated the gas and liquid components of foam as independent fluids flowing in the porous media.

It is well known that when two immiscible phases (e.g. gas and liquid) flow through a porous medium, each phase may be considered to follow separate channels[42]. As the saturation of the immiscible phases changes, the number of channels available for each phase also changes. The effective permeability of each phase is influenced by the percent saturation of that phase. In the presence of the foam, the effective permeability of a porous medium to each phase is considerably reduced as compared to the permeability measured in the absence of foam.

The steam foams can be considered more effective in displacing heavy oil. In general, the higher oil recovery by steam foam flooding is presumably due to the reduction in the oil viscosity and higher foaminess of the surfactants during steam injection compared with air injection. The permeability of porous media is an important factor for foam generation. It is likely that for a porous medium of a given permeability, there may be a specific mean bubble size for maximum fluid displacement.

To gain insight into its flow behaviour, the rheology of foam has been studied[43-45]. There is a general agreement among researchers that foam behaves like a pseudoplastic fluid with high apparent viscosity. The petroleum reservoir fluids are generally considered to exhibit pore channel flow, wherein permeability is a function of saturation only. If foams exhibit a channel flow, its components would move independently and flow through separate and different pore channels. It was observed that the foam continously separates into gas and liquid in the porous medium, continually reforming foam. The investigators[43,44] indicate that the permeability of the foam is not a single function of saturation. This implies that foam does not exhibit pore channel flow.

Foam is considered to be a continuum; the gas and liquid flow at the same rate, and the foam behaves as a single fluid with high apparent viscosity. Foam flow behaviour in a porous medium can not be described merely in terms of a single high viscosity fluid as concluded by Raza[46] on the basis of quality/viscosity measurments of flowing fluids and bulk foam. Minssieux[47] concluded that foam did not flow as a single fluid. He suggested, however, that foam flow can be interpreted using the concept of foam quality as applied to the fluid phase. Foam quality is the ratio of gas volume to the total (gas and liquid) volume of the foam. Despite the difference in these and other experimental findings, it is generally agreed that a low quality foam can and does flow as a single fluid; bubbles are stable, fairly uniform, and small enough to pass through the pore constrictions without breaking.

Foam with large and less stable bubbles is less likely to flow as a single fluid. Mast[40] and Fried[48] deduced that foam is propogated inside a porous medium by the breaking and reforming of foam bubbles. The gas flows as a discontinuous phase while the liquid is transported as a free phase via the film network. Nahid[49] proposed that the gas flow could be treated according to Darcy's law if a correction factor for the gas permeability is used.

The volume of the fluid displaced by gas before its breakthrough at a producing site is termed as fluid displacement efficiency. The fluid displacement efficiency in sandpacks was higher as compared to Berea cores. This indicates that foams are more effective in high permeability zones. It is known that high quality foams (large bubbles) were produced in the high permeability sands which contain uniformly distributed large pores. Mast extended Gardescu's work[50] to obtain equations relating minimum bubble radius with pore dimensions and interfacial properties of the liquid.

The effect of foamer concentration on the nature and type of foam produced in porous media has been studied in some detail by Kander and Schechter[51]. Foamer concentration affects the quality of the foam and the size and nature of the foam bubbles that are produced. The average foam bubble diameter is proportional to foam quality, which increases with concentration. Marsden, et al.[52] observed that the discrepancy between results obtained in different laboratories may have been due to researchers using different foamer concentrations. Fine textured foams are obtained with concentrated foamer solutions, while coarse foams are obtained with dilute foamer solutions. The mobility of gas in the presence of foam decreases with surfactant concentration. The foams are more effective in reducing gas mobility in high permeability as compared to low permeability sandstones.

Bikerman[53] has reviewed various data involving static columns of foam. When gas was bubbled through a surfactant solution, the height of the static column of foam generated was shown to be directly proportional to the gas flow rate. This effect on static foam can be related to flowing foam since the behavior of the latter is affected by bubble texture and size.

The single layer glass bead model is a closer representation of a porous medium. This was used by Sharma[54] to study the foam drive process and Egbogah and Dawe[55] to study the size distribution of oil droplets. Mattax and Kyte[56] used a network of etched capillaries to study fluid distributions under various wettability conditions. Davis and Jones[57] studied the flow of foam in porous media using etched glass micromodels. A study of the multiphase flow of oil and water dispersed in the porous medium was carried out by Bonnet[58] using an etched plastic micromodel.

The thermal stability of several petroleum sulphonates and non-ionic surfactants has been investigated in relation to heavy oil recovery by several workers[59-61]. The stability of the surfactants is reported in terms of the time required for half of the surfactant to decompose (i.e. half life). In order to determine the half-lives for petroleum sulphonates at different temperatures, the activation energy of the surfactants was measured. The precipitation of the sulphonates occurs at the higher temperatures in natural sandstones. It has been reported by Celik et al.[62], Ziegler and Handy[63], that this precipitation appears to occur as a result of an interaction with minerals in the porous media and not due to the direct effects of temperature on the stability of the surfactants.

Several researchers[64-66] have investigated the mechanism and various basic principles involved in CO_2 flooding. It has been reported that at the elevated pressure CO_2 can recover additional oil which is not recoverable by conventional methods. In addition, Supper-critical CO_2 develops multicontact miscibility with various crude oils, resulting in a very efficient oil displacement in porous media. Various efforts have been made by Holm and O'Brien[67] to reduce carbon dioxide channeling by reducing its mobility.

Foam flooding is a method that modifies the flow mechanism by changing the structure of the displacing fluid at the pore level. A critical literature review on foam rheology is given by Haller and Kuntamukkula[68]. The injection of foam-like dispersions or CO_2 foams is a useful method in enhanced oil recovery[69]. This method of decreasing the mobility of a low-viscosity fluid in a porous rock requires the use of a surfactant to stabilize a population of bubble films or lamellae within the pore space of the rock[70]. The degree of thickening achieved apparently depends to some extent on the properties of the rock itself. These properties probably include both the distance scale of the pore space and the wettability, and so can be expected to differ from reservoir to reservoir, as well as to some extent within a given field[71,72].

Falls et al.[73] have investigated the nature of the foam flow by examining the mechanism of foam generation. Beside the investigation on foam mechanism, researchers have tried to find good foaming agents, especially applicable to enhanced oil recovery[74]. Furthermore, due to the harsh reservoir conditions, studies on the compatibility of the surfactant with oilfield brines at reservoir temperatures and pressures have been made[75]. In order to understand the nature and mechanisms of foam flow in the reservoir, some investigators have examined the generation of foam in glass bead packs.

It can be anticipated that all gas-flood projects, as they are presently being carried out, will leave a large fraction of the reservoir oil uncontacted by the injected fluids. This bypassed oil will remain inplace, undisplaced by the injected fluid. Thus, in each current field project, the amount of incremental oil produced by gas flooding could be substantially increased if the uncontacted oil could be reached. The improvement of the vertical and areal distribution of injected fluids through-out the reservoir requires much better methods of sweep and mobility control. The utility of the foams, in general, as mobility control agents has not been extensively tested. In principle they offer a spectrum of fluid mobility behaviour depending on the in-situ foam phase stability.

Many research laboratories are actively involved in investigating the use of surfactants in light and heavy oil recovery. Surfactant and foam flooding may become more widely accepted as more is learned about the mechanism of oil displacement and as these processes are improved and tested in the field. The contact of surfactants with residual oil, connate water, and reservoir rock decreases their efficiency in displacing oil. Therefore, an attempt was made to examine the effect of foam on fluid recovey, oil displacement efficiency, breakthrough time and gas/steam mobility in both sandpacks and sandstones. The experiments for flow through porous medium were designed in such a way so that the surface properties of the surfactant solution and foam characteristics could be correlated with the results of the fluid displacement. The results of this study can be utilized to understand the nature of gas and fluid/oil flow in porous medium.

EXPERIMENTAL

Materials: Several pure and commercial surfactants were used as a foaming agent. All alkyl alcohols (greater than 98% pure) were supplied by Chemical sample company, Columbus, OH, (U.S.A.). All other foaming agents were used as received. Oils (dodecane, tetradecane, hexadecane) were either supplied by BDH (Analar grade) or Chemical Sample Company, Columbus, OH, (U.S.A.). Double distilled water was used in all experiments.

Sand, used as a porous medium, was supplied by AGSCO Corp., Paterson, New Jersey, U.S.A., whereas the Berea sandstones were supplied by Cleveland Quarry, Cleveland, OH, U.S.A.. The size distribution for the sand used was 40-150 micron with average particle size of 95 micron. The sandstone cores were cast in Hysol Tooling Compound (Hi-Co Associates, Orlando, FL) inside PVC pipes. The sandpacks had permeabilities of about 2.5 darcy and porosities of 40%, whereas Berea sandstones had permeabilities of about 275 millidarcy and porosities of 18%. The transducer used for the measurements of pressure across the porous medium was from Validyne Engineering Corp., Northridge, CA, U.S.A.. The recorder was a Heath/Schlumberger Model 225, Heath Co., Benton Harbor, MI, U.S.A.. The water was pumped using a Cheminert metering pump Model EMP-2, Laboratory Data Control, Riviera Beach, FL, U.S.A..

Foaminess of Surfactant Solutions: A glass cylinder with 0.79 in. (2 cm) diameter and 46.5 in. (118 cm) length was used for the measurement of foaminess. The cylinder contained a sieve of 20 to 50 um at the bottom and an outer jacket for water circulation to keep the temperature constant. Foaming solution (20 ml) was poured into the cylinder. The time (5 min.) and pressure (1 psi) for air injection were kept constant for all experiments. The volume of foam generated during a constant time was recorded.

Foam Size Analysis: To determine the bubble size, a plexiglass cell (2" x 2") was constructed for the measurements of foam size. The surfactant solution (10 ml) was placed into the cell, and foams were generated by shaking the cell for a constant time, (e.g. 2 minutes). An average bubble size was determined from the size frequency analysis.

Characterization of Porous Media: Several sandpacks and sandstones (1 ft. long) were used as a porous medium for displacement experiments. The porous medium was flushed with carbon dioxide to replace interstitial air. To determine the pore volume of the porous media, several pore volumes of distilled water were pumped through it. By this method, the trapped gas bubbles in the porous media could be eliminated because carbon dioxide being soluble in water. The water retained in the porous medium was calculated for determining the pore volume. Moreover, water was also pumped through the porous medium at several flow rates to determine the absolute permeability of the porous medium using Darcy's equation. After the porous medium was characterized, surfactant solutions of known surface properties were injected at constant flow rate. This injection of surfactant was followed by gas/steam flow. The pressure difference across the porous medium was measured using a pressure transducer and recorder. The breakthrough time and fluid displacement efficiency were recorded at various temperatures.

THEORETICAL

The fluid displacement experiments were conducted under different conditions of gas injection and types of the porous media. In order to be consistent in the analysis of the results, the flow of the injected gas has been treated according to Darcy's law. Darcy's law for the gas phase flowing in a linear system is given as follows (43).

$$Q_{sc} = \frac{T_{sc} A\ K_g (P_1{}^2 - P_2{}^2)}{2P_{sc}\ T\ Z\ u_g\ L} \tag{1}$$

where Q_{sc} is the constant gas flow rate measured at some standard absolute pressure (P_{sc}) and temperature (T_{sc}), A is the flow cross-sectional area, K_g is the effective gas permeability, T is the flow temperature, L is the length of the porous medium, Z is the gas deviation factor at flow conditions, u_g is the gas viscosity and P_1 and P_2 are the upstream and downstream absolute pressures.

The fluid displacement experiments were carried out at constant temperature and the downstream pressure was kept atmospheric. The pressure drops across the porous medium were always less than 50 psi and thus, the gas deviation factor can be neglected (43). Cumulative volume (q) of the fluid collected at gas breakthrough time (t) was recorded at room temperature and atmospheric pressure. Therefore, the flow rate term Q_{sc} can be replaced by the term q/t, which represents an average flow rate during an experiment until gas breakthrough. Equation (1) can be rewritten as follows.

$$q/t = \frac{K_g\ A\ (P_{11})}{u_g\ L} \tag{2}$$

where

$$P_{11} = \frac{P_1{}^2 - P_2{}^2}{2\ P_{sc}} \tag{3}$$

and

$$K_g = K\ K_{rg} \tag{4}$$

K is the absolute permeability of the porous media and K_{rg} is the relative permeability to gas phase. Equation (2) can be rearranged in various ways to define groups of parameters used in the analysis and interpretation of results.

$$\text{Effective Gas Mobility} = \frac{K_g}{u_g} = \frac{q\ L}{A\ t\ P_{11}} \tag{5}$$

$$\text{Relative Gas Mobility} = \frac{K_{rg}}{u_g} = \frac{q\ L}{K\ A\ t\ P_{11}} \tag{6}$$

TABLE I. SEVERAL COMMERCIAL SURFACTANTS USED AS FOAMING AGENTS

SURFACTANTS	CHEMICAL COMPOSITION
<u>Supplier: Conoco Company</u>	
Alfonic 1012-40	$CH_3(CH_2)_xCH_2(OCH_2CH_2)_nOH$
	where, x = 8-10, n = 2.5
Alfonic 1012-60	where, x = 8-10, n = 5.7
Alfonic 1216-22	where, x = 10-14, n = 1.2
Alfonic 1412-40	where, x = 10-12, n = 3.0
Alfonic 1412-60	where, x = 10-12, n = 7.0
Alfonic 1218-70	where, x = 10-16, n = 10.7
<u>Supplier: GAF Company</u>	
Alipal CD-128	nonadecyldiethoxy ammonium sulphate
Alipal CO-436	nonylphenoxypolyethoxyethanol
<u>Supplier: BASF Wyandotte</u>	
Klearfac AA270	phosphate ester
Klearfac AA420	phosphate ester
<u>Supplier: Rohm & Haas</u>	
Triton X-100	octylphenyl nonylethoxy ethanol
Triton X-165	octylphenyl hexadecylethoxy ethanol
Triton X-2oo	sodium alkylaryl polyether sulphonate
Triton X-301	sodium alkylaryl ether sulphate
Triton X-405	octylphenyl tetracontylethoxy ethanol

TABLE II. FOAMINESS OF VARIOUS SURFACTANTS
(0.1 wt% surfactant in brine, neutral pH, room temperature)

SURFACTANTS	FOAM VOLUME, ml	FOAM HALF-LIFE, min.
Alfonic 1012-40	155	3.5
Alfonic 1012-60	225	5.8
Alfonic 1216-22	140	3.5
Alfonic 1412-40	30	25.7
Alfonic 1412-60	210	7.5
Alfonic 1218-70	195	5.5
Alipal CD-128	270	8.6
Alipal CO-436	230	4.8
Klearfac AA270	225	4.5
Klearfac AA420	165	4.2
Triton X-100	240	4.8
Triton X-165	235	5.0
Triton X-200	80	12.5
Triton X-301	90	4.1
Triton X-405	225	3.6

The effect of the permeability of porous media and pressure drop on flow of a gas in the presence of "foam" was studied. The data obtained were interpreted using equations (5) and (6).

RESULTS AND DISCUSSION

Surfactant flooding for enhanced oil recovery (EOR) involves the use of commercial surfactants due to economic reasons. Limited studies of pure surfactant, sodium dodecyl sulfate (SDS) were conducted in order to compare the results observed using commercial surfactants, as well as to better interpret the experimental findings. The chemical composition of various surfactants is listed in Table-I. The results of foam stability and foam volume for several commercial surfactants at room temperature are shown in Tables II and III. It is evident that the foam volume and foam stability increase with increasing ethoxylated chain length in the given surfactant molecule. This observation can be explained on the basis of the solubility of the surfactant. As the ethoxylated chain length (e.g. OCH_2CH_2) decreases in the surfactant molecule, the solubility of the surfactant also decreases, which results in the decrease of the foaminess. Thus, Alfonic 1216-22 exhibits higher foam stability and produces more foam as compared to Alfonic 1012-40. It appreas that the foam properties such as foaminess, foam stability, bubble size etc. depend on the structure of the surfactants used as a foaming agents for enhanced oil recovery.

As the reservoir temperature is higher than the embient temperature, the foam properties were studied at 50°C. The experimental data for various commercial surfactants are recorded in Tables IV and V. The foamability of the first four Alfonic surfactants (Table-I) at 0.1 wt% concentration is considerabily reduced at 50°C as compared to ambient temperature due to the variation in the dynamic behavior of the surfactants at higher temperatures.

It is concluded that the surfactants, which are persistant to high temperature, can be selected in foam flooding for enhanced oil recovery. Among the surfactants studied, Alfonics 1412-60, 1218-70; Alipals CD-128, CO-436; Klearfacs AA270, AA420; and Tritons X-100, X-165, X-405 produced almost the same foam at 0.1 wt.% concentration both at room and 50°C, while the rest of the surfactants could not produce the same amount of foam at 50°C. For 0.5 wt.% concentration, Alfonics 1012-40, 1012-60, 1412-60, 1218-70; Alipals CD-128, CO-436; Klearfac AA270; Trotons X-165, X-405 produced almost the same amount of foam at room temperature and 50°C. It should be noticed that the foaminess of several surfactants decreased at 0.1 wt.% concentration and 50°C, but remained the same at 0.5 wt.% surfactant concentration both at 50°C and room temperatures. These results indicate that the concentration plays an important role in producing foam at different temperatures.

The photomicrographs of the foams for SDS and Suntech IV are shown in Figure 1. It was observed that the SDS produced smaller bubbles compared to Suntech IV surfactant. This appears to be due to the high surface activity of the SDS in an aqueous medium. The foaminess, bubble size and foam stability as a function of temperature for SDS and Suntech IV surfactants are plotted in Figures 2-4. For both surfactants, foaminess increases with increasing temperature, while bubble size and foam stability (e.g., half life) decrease with increase in temperature. These results demonstrate that the pure and

TABLE III. FOAMINESS OF VARIOUS SURFACTANTS
(0.5 wt% surfactant in brine, neutral pH, room temperature)

SURFACTANTS	FOAM VOLUME, ml	FOAM HALF-LIFE, min.
Alfonic 1012-40	185	4.2
Alfonic 1012-60	265	6.5
Alfonic 1216-22	170	6.2
Alfonic 1412-40	145	4.6
Alfonic 1412-60	240	9.5
Alfonic 1218-70	245	7.0
Alipal CD-128	530	10.2
Alipal CO-436	235	7.5
Klearfac AA270	315	4.5
Klearfac AA420	210	4.2
Triton X-100	295	7.0
Triton X-165	285	6.5
Triton X-200	80	36.0
Triton X-301	55	36.5
Triton X-405	255	6.8

TABLE IV. FOAMINESS OF VARIOUS SURFACTANTS

(0.1 wt% surfactant in brine, neutral pH, 50°C)

SURFACTANTS	FOAM VOLUME, ml	FOAM HALF-LIFE, min.
Alfonic 1012-40	25	3.6
Alfonic 1012-60	75	5.4
Alfonic 1216-22	10	-
Alfonic 1412-40	15	4.5
Alfonic 1412-60	195	4.2
Alfonic 1218-70	200	3.8
Alipal CD-128	325	5.8
Alipal CO-436	235	5.5
Klearfac AA270	230	2.9
Klearfac AA420	185	3.5
Triton X-100	205	3.4
Triton X-165	225	4.0
Triton X-200	70	7.3
Triton X-301	55	3.8
Triton X-405	205	3.8

TABLE V. FOAMINESS OF VARIOUS SURFACTANTS
(0.5 wt% surfactant in brine, neutral pH, 50°C)

SURFACTANTS	FOAM VOLUME, ml	FOAM HALF-LIFE, min.
Alfonic 1012-40	155	2.5
Alfonic 1012-60	230	3.3
Alfonic 1216-22	45	6.5
Alfonic 1412-40	60	6.0
Alfonic 1412-60	235	4.4
Alfonic 1218-70	260	5.6
Alipal CD-128	585	5.4
Alipal Co-436	275	7.5
Klearfac AA270	240	1.8
Klearfac AA420	10	-
Triton X-100	250	2.8
Triton X-165	285	3.6
Triton X-200	125	3.7
Triton X-301	105	2.8
Triton X-405	230	3.5

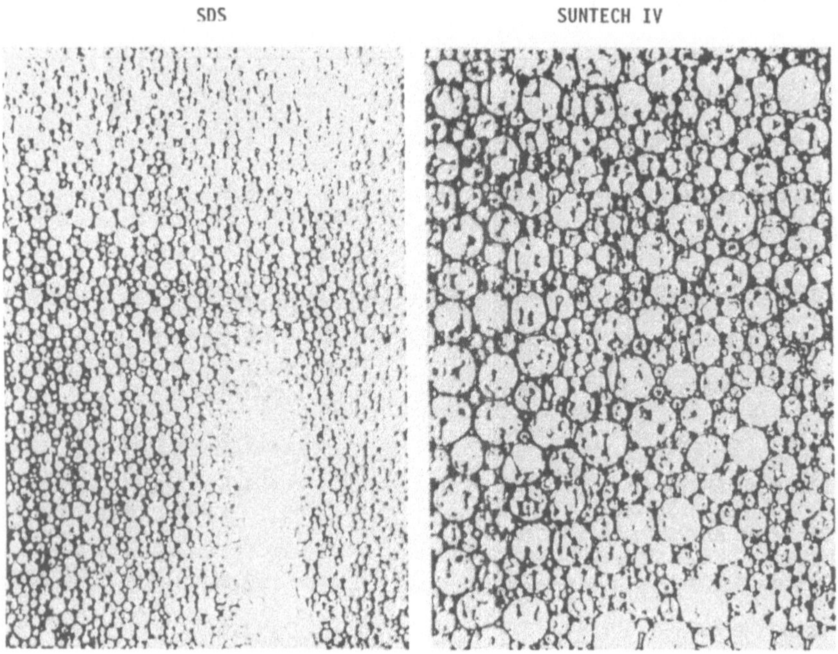

Figure 1. Photomicrographs of the Foams Containing Sodium Dodecyl Sulfate (SDS) and Suntech IV Surfactants.

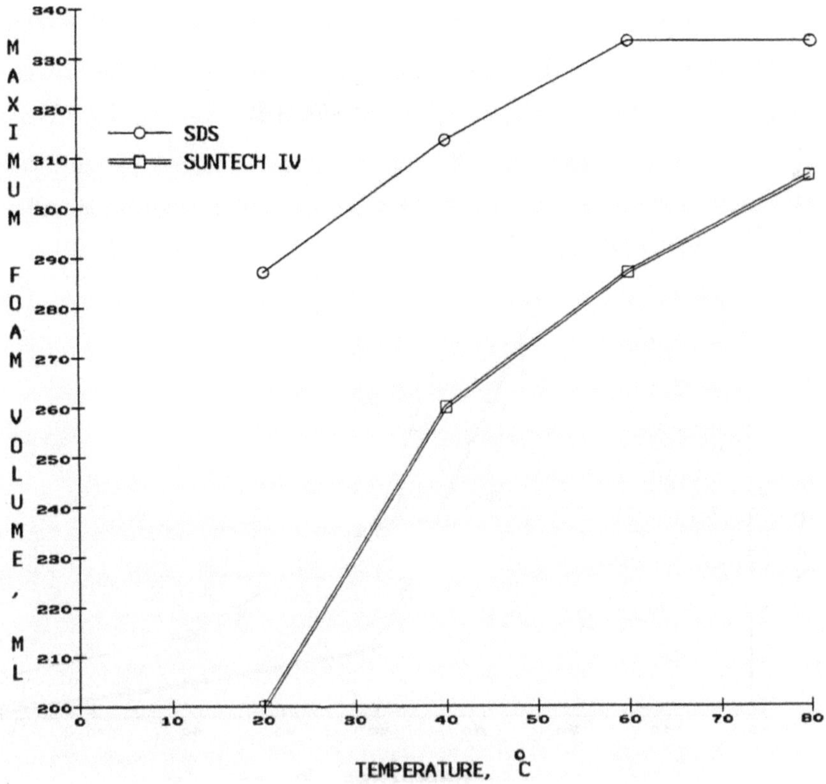

Figure 2. Foaminess as a Function of Temperature for Aqueous Solution of SDS and Suntech IV surfactants.

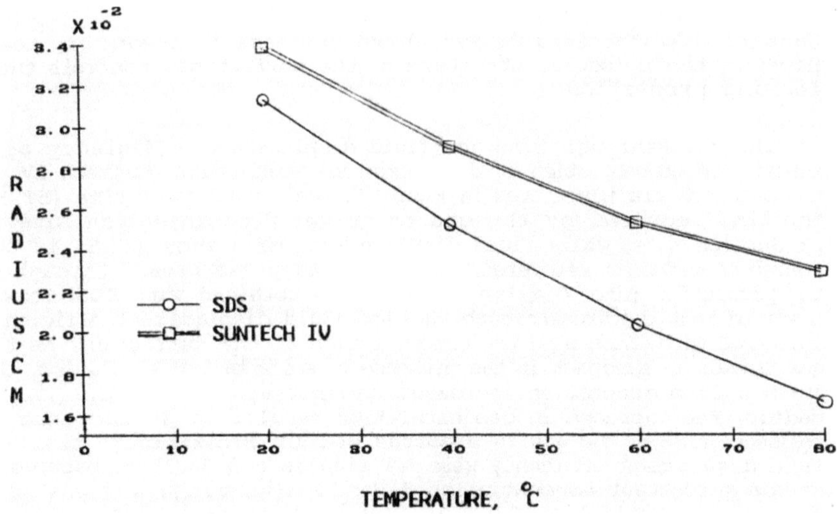

Figure 3. Bubble Size Versus Temperature for SDS and Suntech IV Surfactants.

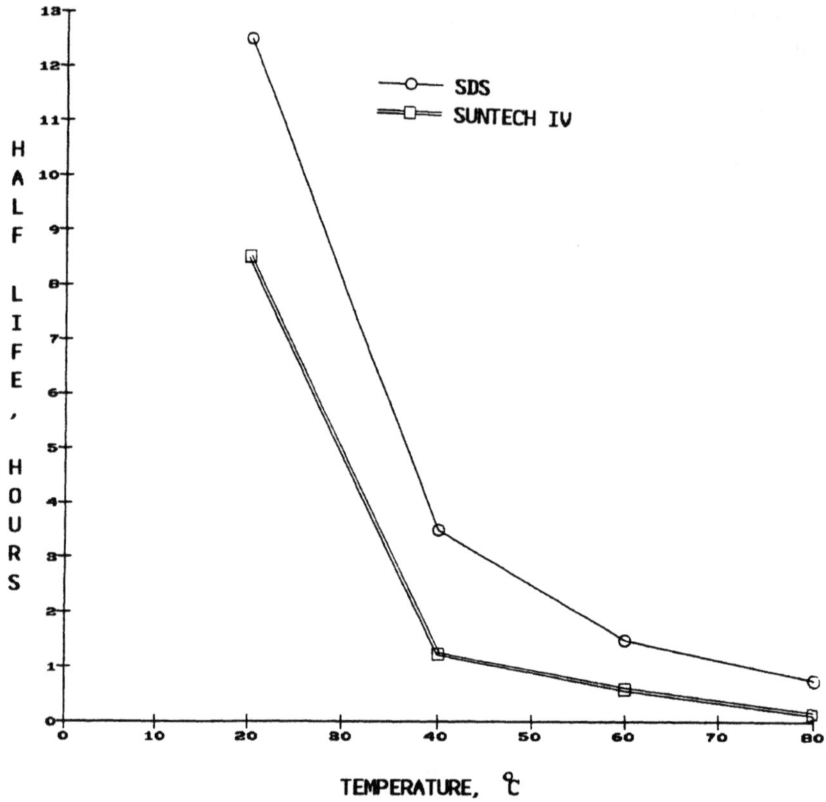

Figure 4. Foam Stability (e.g. Half-Life) Versus Temperature for SDS and Suntech IV Surfactants.

commercial surfactants behave almost the same in producing foam. However, the molecular structure of the surfactants controls their foaming properties.

The breakthrough time and fluid displacement efficiency as a function of concentration of the commercial surfactant, Suntech IV, in the sandpack are illustrated in Figure 5. The breakthrough time (BT) is the time required by the gas to travel from injection site to production site, while fluid displacement efficiency (FDE) is the volume of the fluid recovered before the gas phase breaks through at the production site. Similar results were obtained with SDS. It was observed that the breakthrough time and fluid displacement efficiency increased with increasing the concentration of the surfactant. As the gas travels in sandpack in the presence of surfactant the solution, the in-situ foam generation increases, which trapps gas in the porous medium. The increase in gas saturation results in an increase in residence time of the gas in a porous medium. Breakthrough time and fluid displacement efficiency were 9.5 minutes and 68.5% respectively, at the surfactant concentration 5.0×10^{-4} (wt/vol.), while at the surfactant concentration 1.0×10^{-1} (wt/vol.), these were found to be

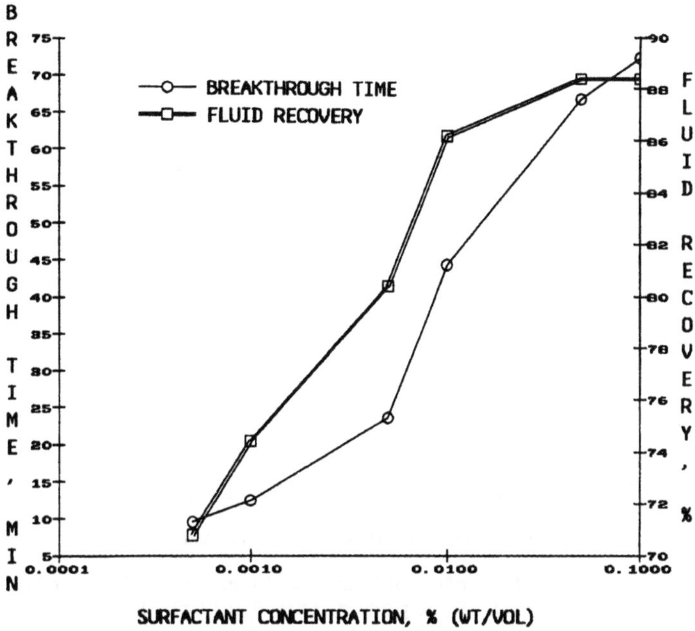

Figure 5. Effect of Surfactant Concentration on Breakthrough Time (BT) and Fluid Recovery (FR) from Sandpacks in the Presence of Suntech IV Foam.

72.2 minutes and 88.4% respectively. At low surfactant concentrations the increase in the BT and FR is slow. Moreover, a gradual increase in the breakthrough time and fluid displacement efficiency were observed above the CMC of the surfactant, while a sharp increase was observed below, but close to the CMC. The trend in the variation of both breakthrough time and fluid displacement efficiency is almost the same with increasing surfactant concentration.

The effect of surfactant concentration on effective air mobility for Suntech IV is shown in Figure 6. It is evident that the effective air mobility is dependent on the surfactant concentration. At low surfactant concentration, the effective air mobility is high, and decreases with increasing concentration. The effective air mobility decreases sharply as the concentration of the surfactant increases up to 2.0×10^{-2} % wt.. Beyond this, a gradual decrease in the effective air mobility is observed. These results show that the foams are effective in reducing the air mobility in the porous media. It was observed that for high fluid displacement, the effective air mobility should be minimal which can be achieved up to some extent by using surfactant concentrations close to the CMC.

Table VI represents the effect of the permeability of porous media (e.g., Berea core) on breakthrough time and relative air mobility at various pressure drops in the presence of Suntech IV. A sharp decrease in the breakthrough time with increasing permeability of the porous

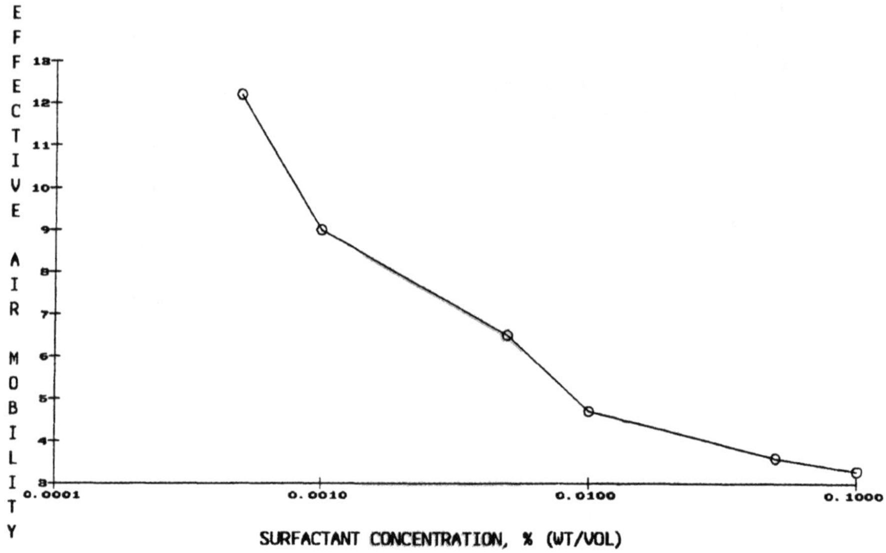

Figure 6. Effect of Surfactant Concentration on Effective Air Mobility in Sandpacks for Suntech IV System.

media (i.e., 50 mD to 200 mD) was observed at the different pressure drops. It is obvious that low permeability restricts the flow of foam through porous media to a greater extent than that of high permeability. Therefore, the resilient interfaces of the bubbles need more time to pass through the media of low permeability as compared to media of high permeability. It was observed, as the permeability is increased, the flow of the foam through porous media is also increased. Therefore, a slight decrease in breakthrough time was noticed in high permeability regions of the porous media. It is clear from Table VI that the relative air mobility decreased (inverse of the relative air mobility increased) with increasing the absolute permeability and pressure drop of the porous media. These results show that the foams are more effective in reducing gas mobility in high permeability as compared to low permeability sandstones. Thus, effectiveness of the foam to recover additional fluid as well as to reduce the gas mobility in porous media can be achieved by increasing the permeability and pressure drop of the porous media.

Figure 7 shows the effect of slug size on breakthrough time and fluid recovery from porous media in the presence of Suntech IV. It is evident from the figure that breakthrough time and fluid displacement efficiency can be correlated as a function of slug size. It was observed that both breakthrough time and fluid displacement efficiency were increased with increasing slug size.

The effect of temperature on fluid displacement efficiency in the presence of SDS and Suntech IV systems is shown in Figure 8. It was

TABLE VI. EFFECT OF PERMEABILITY OF POROUS MEDIA ON BREAKTHROUGH TIME AND RELATIVE AIR MOBILITY AT VARIOUS PRESSURE DROPS IN THE PRESENCE OF SUNTECH IV SURFACTANT SYSTEM.

Permeability	Breakthrough Time (min)			(1/Relative Air Mobility) x 10^{-2} sec.darcy.psi/cc.ft		
(mD)	30 psi	40 psi	50 psi	30 psi	40 psi	50 psi
50	38.4	27.2	22.6	—	—	—
200	14.2	10.5	8.2	6.0	6.5	7.6
400	10.0	8.0	5.2	6.2	6.8	7.8
600	7.8	6.2	3.8	6.8	7.2	8.5
800	6.4	5.6	3.2	7.0	7.8	9.2
1000	5.4	4.4	2.8	7.4	8.2	10.4
1200	4.2	3.8	2.6	7.6	8.8	11.2

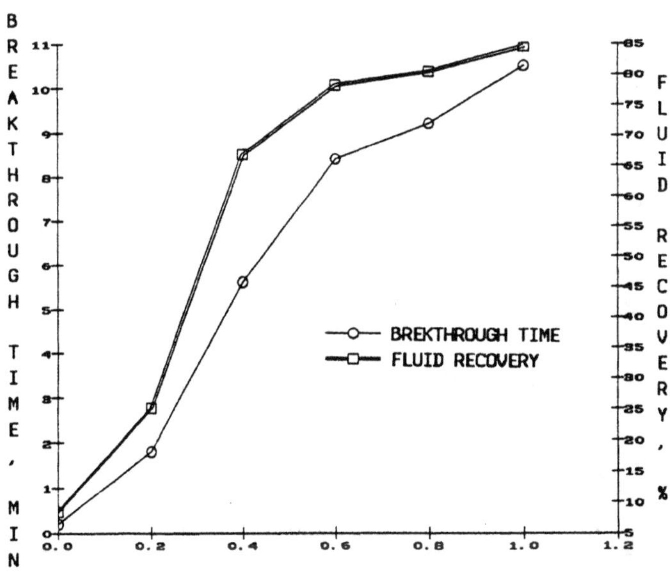

Figure 7. Effect of Slug Size on Breakthrough Time (BT) and Fluid Recovery from Sandpacks for Suntech IV Foam System.

observed that the fluid displacement efficiency increased with temperature for both surfactant systems. A steep increase in fluid recovery was observed in the temperature range of 20-40°C. Beyond this temperature, the fluid recovery increased gradually. The results indicate that upon increasing temperature, the surface tension as well as bubble size decreased which, in turn, increased the fluid displacement efficiency. As the temperature is raised, the kinetic energy of the molecules increases, resulting in a decrease in attractive forces between the molecules, reducing surface tension as well as bubble size. However, the ability of both pure and commercial foaming agents to produce foam was increased with raising temperature.

Figure 9 shows the effective air mobility as a function of temperature at various pressure drops in the presence of SDS. The effective air mobility in the presence of surfactant solution decreased with increasing temperature. However, a sharp decrease in effective air mobility was observed in the temperature range 20-40°C. Beyond this temperature range, the effective air mobility decreased gradually. The trend in the variation of effective air mobility as a funtion of temperature is the same for different pressure drops. It was noticed that the foaming tendency of surfactants increased with increasing temperature and pressure drop. The effectiveness of the foam to recover additional fluid as well as to reduce gas mobility in porous media, therefore, increased with increasing temperature and pressure drops.

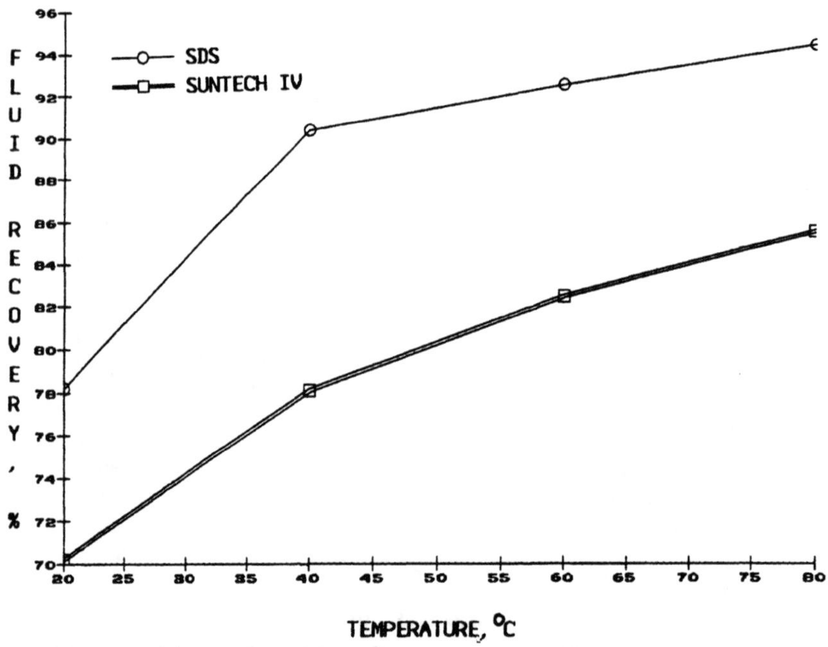

Figure 8. Effect of Temperature on Fluid Recovery from Sandpacks for SDS and Suntech IV Foam Systems.

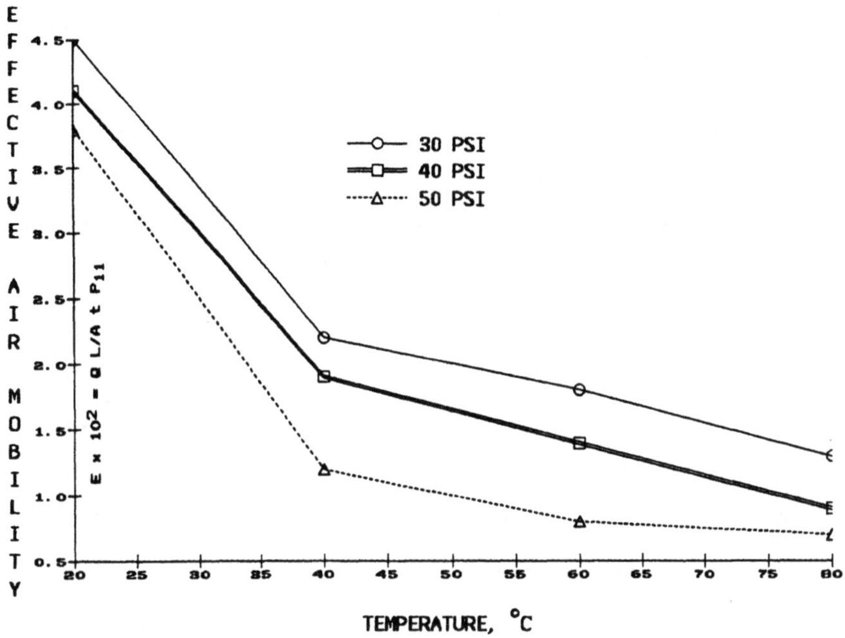

Figure 9. Effect of Temperature on Effective Air Mobility at Various Pressure Drops for SDS System.

These results also indicate that high foam stability is not required for reducing gas mobility or for higher fluid recovery.

Table VII represents the variation in breakthrough time and fluid displacement efficiency as a function of applied air pressure with and without foaming agents in sandpacks. To keep the breakthrough time within measurable limits, the fluid displacement efficiency and breakthrough time were measured between 2.0 and 20.0 psi. A small decrease in fluid displacement efficiency with increasing applied pressure was observed in the absence of foam. A sharp decrease in the breakthrough time was observed in the pressure range 2-4 psi, and beyond this, a gradual decrease occured. The breakthrough time and fluid displacement efficiency were also measured in the presence of foam at different applied pressures. An increase in fluid displacement efficiency was observed at low pressures. The maximum fluid displacement occured at 8 psi. Beyond this pressure, the fluid displacement efficiency decreased gradually. The trend in the breakthrough time was observed to be the same in both (presence and absence of foaming agent) cases. In the presence of foam, the breakthrough time decreased sharply in the pressure range 2-4 psi and gradually decreased beyond this pressure range.

The effect of oil viscosity on fluid displacement efficiency for both pure and commercial surfactants is shown in Figure 10. It was observed that the oil viscosity has an effect on the fluid displacement

TABLE VII. EFFECT OF PRESSURE DROPS ON BREAKTHROUGH TIME AND FLUID RECOVERY FROM SANDPACKS WITH AND WITHOUT FOAMING AGENTS

Pressure Drops Psi	Breakthrough Time (min) with foam	Breakthrough Time (min) without foam	Fluid Recovery % (OFIP) with foam	Fluid Recovery % (OFIP) without foam
2.0	210.0	2.80	68.2	7.8
4.0	80.0	1.40	70.0	7.5
8.0	55.0	1.15	78.4	7.2
12.0	28.0	0.50	75.8	6.6
16.0	12.0	0.20	72.6	6.3
20.0	6.0	0.18	66.5	6.1

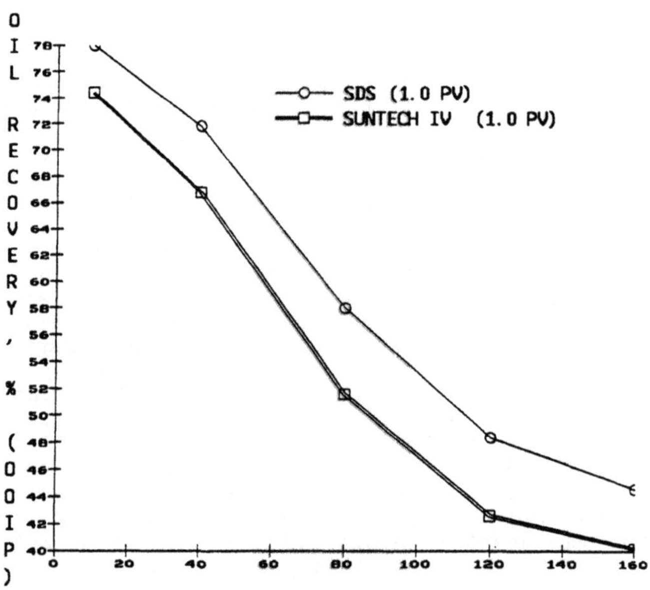

Figure 10. Effect of Oil Viscosity on Oil Recovery from Porous Media in the Presence of SDS and Suntech IV Foams.

efficiency. The curves in Figure 10 clearly show that the oil displacement efficiency is decreased with increasing oil viscosity. The decrease in oil recovery is sharp below 120 cps viscosity of oil in the presence of SDS and Suntech IV surfactant systems. Beyond this, a gradual decrease in oil recovery was observed for pure and commercial surfactants. These results indicate that the behaviour of pure and commercial surfactants in a porous medium is almost the same under these expermental conditions. More viscous oils are recovered in lower amount as compared to the less viscous oils.

ACKNOWLEDGEMENT

The author would like to thanks the authorities of PSE, INC., Kingsport, TN (U.S.A.) for providing facilities to carry out a part of the work presented here.

REFERENCES

1. Bond, D.C. and Holbrook, O.C.; Patent No.2,866,507, Gas drive oil recovery process (1958).

2. Jacobs, W.L. and Bernard, G.G.; U.S. Patent No.3,330,346 (1965).

3. Craig, F.F.,Jr. and Lummus, J.L.; U.S. Patent No.3, 185,634 (1965).

4. Albrecht, R.A. and Marsden, S.S.,Jr.; Soc. Pet. Eng. J., 10, 51-55 (1970).

5. Bernard, G.G. and Holm, L.W.; Soc. Pet. Eng. J., 10, 9-16 (1970).

6. Belkin, H., Holbook, S. and Kuntamukkula, M.; 1st Annual Report, New Mexico State University, Las Cruces, NM., 1-66 (1979).

7. Sharma, M.K. and Shah, D.O.; J. Am. Oil Chem. Soc., 61(3), 585-590 (1984).

8. Sharma, M.K., Shah, D.O. and Brigham, W.E.;Annu. Convent. DOE Contractor, Enhance oil recovery project, San Francisco,Calf., July 27-29, 74-84 (1982 b).

9. Sharma, M.K. and Shah, D.O.; 184th Natl. Meet., Am. Chem. Soc., Washington,D.C.,(U.S.A.),August 21-26 (1982).

10. Mannheimer, R.J. and Schechter, R.S.; J. Colloid Interface Sci., 25, 434 (1967).

11. Mannheimer, R.J. and Schechter, R.S.; J. Colloid Interface Sci., 27, 324 (1968).

12. Mannheimer, R.J. and Schechter, R.S.; J. Colloid Interface Sci., 32, 195 (1970).

13. Klevens, H.B.; J. Am. Chem. Soc., 72, 3780-3785 (1950)

14. Elson, T.D. and Marsden, S.S.; Calif. Reg. Meet.,SPE AIME San Francisco, Calif., April 12-14, 1-7,SPE 7116 (1978).

15. Corrin, M.L. and Harkins, W.D.; J. Am. Chem. Soc., 69, 683-688 (1947).

16. Ginn, M.E., Kinney, F.B. and Harris, J.C.; J. Am. Oil Chem. Soc., 37, 183-187 (1960).

17. Adam, N.K.; The Physics and Chemistry of surfaces, Oxford University Press, London (1941).

18. Kitchener, J.A. and Cooper, C.F., The Chemical Society of London, 8 (1959).

19. Puddington, I.E.; "Foams", Pulp and Paper of Canada, 58,243 (1957).

20. Burcik, E.J.; J. Colloid Science, 5, 411 (1950).

21. Miles, G.D. and Ross, J.; J. Physical Chemistry, 48, 280 (1944).

22. Wason, D.T., Nikolov, A.D., Huang, D.D. and Edwards,D.A., In Surfactant-Based Mobility Control" Ed. by Smith, D.H., ACS Symposium Series, 373, Am. Chem. Soc., Washington,D.C. (1988).

23. Hu, P.C., Tuvell, M.E. and Bonner, G.A.; Paper SPE/DOE 12660, presented at the 4th Symposium on Enhanced Oil Recovery, April 15-18 (1984).

24. Dunning, N., Eakin, J.L. and Walker, C.L.; Monograph No.11, U.S.Bureau Of mines, 38-47 (1961)

25. Pletnev, M.Y. and Trapeznikov, A.A.; Presented at the 2nd All-Union conference [in U.S.S.R. (in russian)] Ed. by Shebekino, 33-52 (1979).

26. Roberds, K., Axberg, C., Osterlund, R.; J. Colloid Interface Sci., 62, 264-272 (1977).

27. Prince, A.; Seminar presented at the Illinois Institute of Technology, June (1985).

28. Reiss-Husson, F. and Luzzate, V.; J. Phys. Chem., 68, 3504-3511 (1964).

29. Efremov, I.F.; In "Surface and Colloid Science", Ed. by Matijevic, E., Wiley Inter. Sci., 85, N.Y. (1976).

30. Roof, J.G.; Soc. Pet. Eng. J., 85-90, March (1970).

31. Mohanty, K.K., Davies, H.T. and Scriven, L.E.; SPE 9406 presented at the 55th Annu. Fall Meeting of SPE-AIME, Dallas Texas, Sept. 21-24 (1980)

32. Wardlaw, N.C. and Cassan, J.P.; Bulletin of Canadian Petroleum Geology, 27, No.3, 402-404, Sept. (1979).

33. Goodrich, J.H. and Watson, J.A.; In "Target Reservoirs For Carbon dioxide miscible Flooding", Final Report, Vol.2 DOE/MC/08341-17 (1980).

34. Bernard, G.G.; Proc. Monthly Vol.27, No.1, 18 (1963).

35. Bennett, G.S.; M.S. Thesis in Petroleum and Natural Gas Eng., Pennsylvania State University (1963).

36. Kolb, G.E.; M.S. Thesis, Pet. & Nat. Gas Eng., Penn. State Univ. (Dec., 1964).

37. Bernard, G.G.; U.S. Patent No. 3, 330, 351 (1967).

38. Eakin, J.L. and Eckard, W.E.; Pet. Eng., 71, July (1966).

39. Nasibor, N.K.; Nefterom Delo., No.5, 23 (1970).

40. Mast, R.F.; SPE 3997 Presented at SPE-AIME 47th Annual Fall Meeting, San Antonio, Texas, 8-11 October (1972).

41. Marsden, S.S.,Jr.; Class notes for petroleum engineering 284, Stanford University (1979)

42. Caudle, B.H. and Dyes, A.B.; Trans AIME, 213, 281 (1958).

43. Raza, S.H. and Marsden, S.S.; Soc. Pet. Eng. J., 359-368 (Dec. 1967).

44. David, A. and Marsden, S.S.; SPE 2544 Presented at the 44th Annual Fall Meeting of SPE-AIME, Denver, Colo., September 28 October 1 (1969).

45. Holbrook, S.T., John, T.P. and Hsu, W.; SPE/DOE 9809, Presented at the SPE/DOE second joint symposium on EOR of the SPE, Tulsa, OK, April 5-8 (1981).

46. Raza, S.H.; Soc. Pet. Eng. J., 328-336 (Dec. 1970).

47. Minssieux, L.; J. Pet. Tech., 100-107 (Jan. 1974).

48. Fried, A.N.; U.S. Bureau of mines, R.I. 5866 (1961).

49. Nahid, B.h.; Ph.D. Thesis, University of Southern California Los Angeles (1971).

50. Gardescu, I.I.; Trans. AIME, 86, 351 (1930).

51. Kander, M. and Schechter, R.S.; SPE 6200, Presented at the 51st Annual Fall Meeting of SPE-AIME, New Orleans, LA., October 3-6 (1976).

52. Marsden, S.S.,Jr., Elson, T. and Guppy, K.; TR-3, Stanford University Petroleum Research Institute, Stanford, California, November (1977).

53. Bikerman, J.J.; Foams, Springer-Verlag New York Inc. (1973)

54. Sharma, S.K.; M.S. Report, Stanford University, (Dec. 1965)

55. Egbogah, E.O. and Dawe, R.A.; Bulletin of Canadian Petroleum Geology, 28, No.2, 200 (1980).

56. Mattax, C.C. and Kyte, J.R.; Oil and Gas J., 115 (Oct. 16, 1961).

57. Davis, J.A. and Jones, S.C.; J. Pet. Tech., 1415 (Dec. 1968).

58. Bonnet, J.; Ph.D. Thesis, Toulouse University, Toulouse, France (1978).

59. Handy, L.L., Amaefule, J.O., Ziegler, V.M. and Ershaghi, I.; Inst. Symp. on Oil Field and Geothermal Chemistry, Houston, Texas, January 22-24 , 61, SPE 7867 (1979).

60. Handy, L.L.; In Proceedings, European Symposium on Enhanced Oil Recovery, Bournemouth, England, September 21-23. Elsevier, Lausanne, 149 (1981).

61. Isaacs, E.E., Prowse, D.R. and Rankin, J.P.; 32nd Annu. Tech. Meet., Pet. Soc. CIM, Calgary, Alta., May 3-6, 1 (1981)

62. Celik, M., Goyal, A., Manev, E. and Somasundaram, P.; 54th Annu. Tech. Conf., SPE-AIME, Las Vegas, Nev., September 23-26, SPE 8263 (1979).

63. Ziegler, V.M. and Handy, L.L.; Soc. Pet. Eng. J., 21(2), 218 (1981).

64. Holm, L.W.; Trans. AIME, 216, 225 (1959).

65. Holm, L.W.; Prod. Mon., 27(9), 6 (1963).

66. Holm, L.W. and Josendal, V.A.; J. Pet. Technol., 26(12), 1427 (1974).

67. Holm, L.W. and O'Brien, L.J.; J. Pet. Technol., 23(4), 431 (1971).

68. Heller, J.P. and Kuntamukkula, M.S.; Ind. Eng. Chem. Res., 26, 318 (1987).

69. Heller, J.P., lien, C.L. and Kuntamukkula, M.S.; Soc. Pet. Eng. J., 25, 603 (1985).

70. Radke, C.J. and Ransohoff, T.C.; Presented in part at the 61st SPE Annual Technical Conference and exhibition, New Orleans, LA, October (1986).

71. Heller, J.P.; Presented at the SPE/DOE 4th Symposium on Enhanced Oil Recovery, Tulsa, OK, April (1984).

72. Heller, J.P., Boone, D.A. and Watts, R.J.; Presented at the 60th Annual Technical Conference and Exhibition of SPE, Las Vegas, NV, September (1985).

73. Falls, A.H. et al.; Presented at the SPE/DOE 5th Symposium on Enhanced Oil Recovery, April (1986).

74. Borchardt, J.K.; Presented at the SPE International Symposium on Oil Field Chemistry, San Antonio, TX, February (1987).

75. Maini, B.B., Ma,V.; J. Can. Pet. Tech., 25, 65 (1986).

DETECTION OF A NEW EFFECT AS A RESULT OF POLYMER BEHAVIOR IN ALKALINE MEDIA

I. Metin Mihcakan and Craig W. Van Kirk

Colorado School of Mines
Golden, Colorado

An unexpected Residual Oil Saturation (ROS) ring was observed after the application of the Polymer Augmented Alkaline Flooding (PAAF) process in linear and radial Berea sandstone cores. No information about the previous observations of this effect has been found in the literature. Such an effect was not observed in cores flooded with either alkaline solution or polymer solution alone. Thus, the formation of the ROS ring was attributed to the interaction between the alkaline/polymer blend solution and chasing fresh polymer solution at their transitional interface.

The visible ROS ring was located next to the zone which was totally cleaned out of oil near the injection sand face. Thickness of the ROS ring appears to be a direct function of the slug size of the alkaline/polymer blend solution.

A hypothesis is developed to explain the formation mechanism of the ROS ring by the polymer rheology in high pH caustic media influenced by the microscopic heterogeneity of porous medium. Moreover, the Reciprocal Relative Mobility (RRM) data were utilized for studying the propagation of the transitional interface between the leading blend slug and the chasing fresh polymer slug across the cores of both geometries.

Particle Technology and Surface Phenomena in Minerals and Petroleum
Edited by M.K. Sharma and G.D. Sharma, Plenum Press, New York, 1991

INTRODUCTION

Alkaline flooding (AF) is a chemical process used in the Petroleum Industry to remove oil from pore walls and to mobilize the mechanically entrapped oil globules or ganglia in porous media. Among all the enhanced oil recovery (EOR) processes, the AF is one of the most complicated and least understood due to its complex oil recovery mechanisms.

However, all the researchers agree on one theme that the crude oil must contain acidic components to be able to react favorably in an alkaline medium. These acidic components, e.g., carboxylic acids, phenols, porphyrins, etc., are responsible for the surface activity which is one of the major factors controlling the efficiency of the AF process(1).

A number of different alkaline agents, e.g., sodium hydroxide, sodium carbonate, sodium orthosilicate, ammonium hydroxide, etc., have been considered and added to flood-waters throughout the world, since the early 1900's(2-4). In all these field applications two major drawbacks of the AF process have arisen: (1) poor displacement efficiency caused by excessive alkaline consumption, and (2) poor sweep efficiency and fingering caused by low alkaline solution viscosity (unfavorable mobility ratio), in reservoirs(4,5).

Once the alkaline solution is injected into the reservoir, two types of distinct interactions take place to consume the available alkalinity: (1) interactions between the alkaline solution and reservoir rock(6,7), and (2) interactions between the alkaline solution and reservoir fluids(8).

The reactions between the alkaline solution and reservoir oil generate the complex oil recovery mechanisms of AF postulated to date(9): (1) in-situ surfactant generation by neutralization (saponification), (2) reduction of the interfacial tension (IFT) at the oil-water interface, (3) temporary wettability alteration, (4) emulsification with entrainment, (5) emulsification with entrapment, (6) emulsification with coalescence, (7) oil phase swelling, and (8) breaking out the rigid films. Thus, the consumption of alkalinity through the reactions of alkaline solution with reservoir waters and rock constituents has been accepted as the most important disadvantage of the AF process(3,4).

It has been well known since the early 1960's that the

addition of polymers increase the viscosity of floodwaters, so that better mobility control and higher volumetric sweep efficiencies can be achieved. It is also well known that polymers are not capable of improving the displacement efficiency by removing oil from the rock surface and by mobilizing the entrapped oil globules. Instead, the polymers added to floodwaters accelerate the recovery of oil which would otherwise be produced after infinite water throughput in conventional waterflooding[10].

Some researchers, therefore, came up with the idea of utilizing the synergistic effect of the combined favorable characteristics of polymer flooding (PF) and alkaline flooding (AF) to recover additional oil more efficiently. The simultaneous application of these two chemical processes in an appropriately chosen field was not a problem, because the required ranges for crude oil API gravity, reservoir temperature, and reservoir water salinity were about the same for both processes.

Consequently, such an idea led to the development of a new EOR process called the "first version of Polymer Augmented Alkaline Flooding" (PAAF). In the application of this process an alkaline solution slug is injected to mobilize the residual oil, which is the oil mechanically entrapped and/or sticking on the pore walls. The alkaline solution slug is chased by a fresh polymer slug, which provides the improved mobility control and volumetric sweep efficiency to enhance oil recovery[11].

During the early 1980's, soon after the first version of the PAAF process was introduced, a second version was proposed for the PAAF process, which incorporates a single blend slug of alkaline and polymer chased by a fresh polymer slug. The main idea behind the second version of the PAAF process is to employ the capability of polymer adsorption on the rock surface to mitigate the alkaline consumption[12-14].

In other words, the polymer molecules in the blend solution slug would adsorb and coat the water-wet portions of the pore walls, so that the alkaline agent would have only limited sites for the rock dissolution and ion exchange reactions. Hence, the saved useful alkalinity could be used for the reactions with the organic acids in the crude oil in order to mobilize the residual oil. Moreover, as in the first version of PAAF, chasing fresh polymer slug would provide the mobility control and the efficient volumetric sweep.

The second version of the PAAF process recovers approximately the same amount, if not a little more oil than the first version of PAAF, but with much less alkaline agent consumption and in a shorter time frame. It is also reported in the literature that the second version of the PAAF process is superior to either the AF or PF processes alone in increasing the oil recovery.

The EOR processes applied in this present study are the conventional AF, PF, and the second version of PAAF.

EFFECT OF SALTS ON POLYMER SOLUTION VISCOSITY

Among the polymers frequently used by the petroleum industry only the synthetic polymers, such as polyethylene oxides, polyacrylamides, etc., are affected adversly and significantly by the salts present in the solution[15,16]. Since the polyacrylamide polymers are the major concern in this present study, the effect of salts on their rheological behavior will be emphasized and evaluated.

If the molecules of a polyacrylamide polymer are electrically uncharged, then the polymer is called either unhydrolyzed or non-ionic. In this case the polymer molecules are usually loose and roughly spherical coils, while showing a linear-chain configuration, in any solvent. Thus the polymer solution would exhibit a high viscosity[17].

Electrically charged (hydrolyzed) polyacrylamide molecules, on the other hand, tend to expand, similar to the unhydrolyzed polymer molecules, only in good solvents which must be free of salt, such as distilled water. Therefore, a high viscosity would be offered by the polymer solution[17].

Once any type of salt is added to a partially hydrolyzed polyacrylamide (PHPA) solution, the ions will generate an adverse effect on the viscosity of the polymer solution. This adverse effect can be explained as follows[18-20]:

There are some self-repelling electrical charges within the molecules of ionic (PHPA) polymers. In a good solvent these repelling forces are strongest, thus the polymer molecule chains are urged to expand in a rod-like or cylindrical configuration. In other words, the negative charges on the hydrolyzed carboxyl groups (branches of the molecular chain) make each branch repel the others, so that the whole molecular chain would uncoil. These uncoiled and expanded molecular strings entangle with each other to exhibit a high solution viscosity.

When a salt is added into the PHPA solution, monovalent or divalent cations would react with the hydrolyzed carboxyl groups to neutralize the electrical charges on the polymer molecules. Hence, the repelling forces between the branches of the molecules will be either lessened or diminished. At this stage, the polymer molecular chains coil up into rough spherical, tightly wound coils exposing minimum surface to the solvent. This change in shape and size of the polymer molecules, in turn, results in a lower solution viscosity, which remains practically constant at higher salt concentrations[21].

EFFECT OF CAUSTIC ON POLYMER SOLUTION VISCOSITY

A number of technical papers and documents, disclosing the observations on polyacrylamide behavior in caustic environment, were investigated during this study. It was found that almost all the researchers declare unanimously that the viscosity of a polyacrylamide solution increases as the caustic soda (NaOH) concentration in the solution increases, (increasing pH). Additionally, the increase in viscosity levels off at very high pH values (approximately above 11), as in the case of very low pH values (approximately below 6)[21].

On the other hand, only Szabo[22] investigated the long term viscosity behavior of the polyacrylamide polymer in caustic solutions of constant caustic concentration. Based on Figure 1, which is a reproduction from Szabo's paper, the stability of polyacrylamide polymers in caustic solutions of constant caustic concentration can be described as follows:

When an unhydrolyzed polyacrylamide polymer is dissolved in a caustic solution, a reaction starts immediately between the alkaline agent (NaOH) and the amide (NH_2) groups of the polymer molecules. This reaction, which is nothing different from the hydrolyzation process (hydrolysis) of the polyacrylamide molecules, progresses continuously either for several hours or several days, depending on the nature of the polymer used.

During the hydrolyzation process the viscosity of the polyacrylamide solution starts increasing at a high rate, until the viscosity reaches a maximum. At this moment, when the maximum viscosity is reached, the polyacrylamide molecules have already been partially hydrolyzed to a

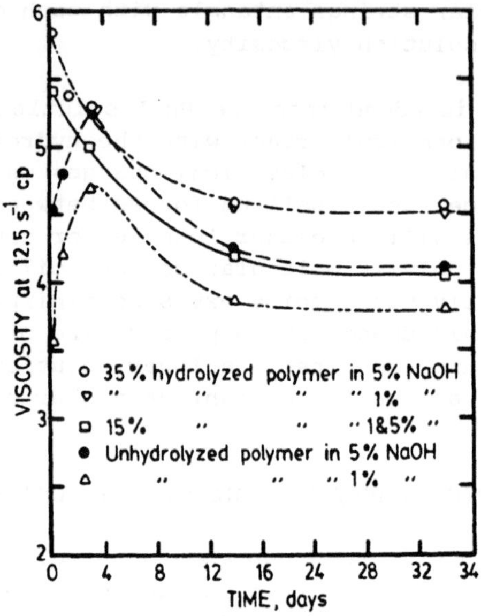

Figure 1. The effect of initial hydrolysis on the time dependent behavior of polyacrylamide polymers in caustic media (after Szabo[22]).

certain degree, which is generally at or little less than 35-percent.

After the viscosity reaches its peak value, the hydrolysis would still be continuing towards a higher degree of hydrolyzation, which cannot contribute much to the viscosity buildup anymore. Thus, the rate of viscosity increase will be diminishing rapidly. In the mean time, the excess sodium ions (Na^+) from the caustic soda (NaOH) would start acting as if they were the monovalent cations from the sodium chloride (NaCl) dissolved in the solution. Such a behavior may be called the "pseudo salt effect."

Such a pseudo salt effect overcomes the diminishing rate of viscosity buildup, and the solution viscosity starts decreasing at a rate which is less than the initial high rate of viscosity increase. Eventually, both the hydrolysis and the viscosity reduction will level off and become constant in time.

Now assume that the same procedure given above is repeated with a polyacrylamide polymer, which is already partially hydrolyzed to a degree of 15 to 35 percent. In this case, the viscosity behavior of the PHPA polymer in caustic solution will not be very different from the repeat of the viscosity reduction part of the behavior exhibited by the unhydrolyzed polyacrylamide polymer in caustic solution.

At this point, it should be noted that the ranges of both time and viscosity scales in Figure 1 may vary depending on the nature of polyacrylamide polymer used, the shear rate at which the viscosity values were determined, the concentration of salts, if there are any, and caustic in solution.

POLYMER RETENTION IN POROUS MEDIA

There are two mechanisms which cause the retention of some fraction of a PHPA polymer injected into the porous medium: (1) the adsorption of polymer molecules on the pore walls, and (2) the mechanical entrapment of polymer molecules (see Figure 2).

The PHPA polymer molecules can be adsorbed on both silica and clay minerals, unless they are coated with oil[23-25]. When the concentration of polymer molecules in solution is low, the coverage on the pore walls is imperfect, and thus the monolayer adsorption occurs. In the case of high polymer concentrations, however, a good coating on the pore walls by the multilayer adsorption is pronounced[25]. Generally, the hydrodynamic thickness of the adsorbed polymer layer is slightly greater than twice the gyration radius of the polymer macromolecule in solution[26].

Since the presence of salts urge the polymer molecules to coil up, the amount of polymer in the adsorbed layer increases with increasing salt concentration[25]. In addition, a higher degree of polymer hydrolyzation leads to a greater polymer adsorption on the pore walls[23]. Thus, these polymer molecules in caustic media would exhibit a high magnitude of adsorption.

When adsorption takes place, the injected polymer solution moves ahead while loosing its polymer content continuously[17]. Once the polymer adsorption on the walls of high permeability flow paths is completed, the effective hydraulic radii and the permeability in these paths are reduced. Hence, some or all of the flow of new incoming

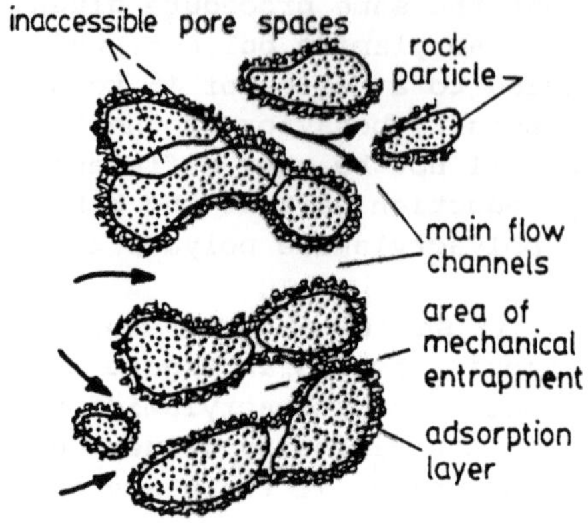

Figure 2. Schematic illustration of polymer retention mechanism in steady-state polymer flow (after Szabo[25]).

polymer solution should be diverted into the other flow paths, which are less permeable yet large enough to allow the flow of macromolecules.

Polymers are retained in the porous media also by the entrapment of the macromolecules mechanically at the small pores and pore constrictions[(23)]. As the first polymer molecule enters the pore, it does not necessarily plug the pore, but it can be adsorbed on the walls of a pore constriction. Another newly arrived polymer molecule may entangle with the previously adsorbed or entrapped polymer molecule. Other incoming new polymer molecules may also accumulate on or around the previous ones. Eventually, there will be enough polymer accumulation to plug the pore, and there will not be any outflow of polymer molecules from this pore. At this moment, the mechanical entrapment of the polymer molecules is completed and the flow is diverted into other available flow paths[(25)].

EXPERIMENTAL DETAILS

In this study the AF, PF, and the second version of PAAF processes were applied in 14 corefloods conducted in linear and radial Berea sandstone cores. The terms linear and radial indicate the flow geometries in the cores during the experiments. The cores were given code names as follows:

SL : Short-Linear plug
LL : Long-Linear plug
SR : Small-Radial disk
LR : Large-Radial disk.

Each linear core plug had a corresponding radial disk, thus they formed a couple. Each core couple was identified by a number added after the code name, such as SL1 and SR1. As shown in Table I, each core couple was chosen so that they would exhibit similar pore volumes (PV), porosities, and absolute permeabilities.

The cores were fired at temperatures gradually increasing to a maximum of 454°C (850°F), in order to neutralize the expandable clays present in the cores. Although a radial coreholder was utilized for the radial cores, the linear cores were cast in epoxy or lucite, which served as coreholders.

INJECTION FLUIDS

The fluids used for core saturation and injection were prepared as explained below. The properties of these fluids are shown in Table II. Note that all the viscosity measurements were made on a Brookfield viscosimeter at 6-RPM (6.57 s^{-1}) to preserve the consistency among the measured values.

Crude Oil

The crude oil used in the experiments was a mixture of 75-percent natural acidic crude oil and 25-percent n-decane, by volume. Therefore, the originally high viscosity of the natural acidic crude oil was decreased to a desired level of approximately 30-cP. Consequently, the injectivity loss, pore plugging and emulsion formation problems could be avoided. The natural acidic crude oil was from the Newcastle formation in the Powder River Basin in Wyoming. The crude oil prepared was then filtered through 1.2 and 0.8-μm Millipore filter papers.

Produced Water

The produced water (PW), which was used for core saturation, was a synthetic simulation of an actual

Table I. Core Properties

Core	Length (cm)	Height (cm)	Diameter (cm)	Pore Volume (cc)	Porosity (%)	Permeability (μm^2)
SL1	15.173	-	5.08	65.37	21.3	1.207
SL2	14.473	-	5.08	64.28	21.9	0.725
SL3	14.346	-	5.08	64.07	22.0	1.145
SL4	14.199	-	5.08	63.80	22.2	1.125
SL5	14.521	-	5.08	65.57	22.3	1.195
LL1	45.949	-	5.08	212.0	22.8	0.991
LL2	45.949	-	5.08	205.0	22.0	0.839
LR1	-	5.043	14.850	197.0	22.6	0.772
LR2	-	5.075	14.859	200.1	22.8	0.780
SR1	-	5.085	8.237	59.0	21.8	0.888
SR2	-	5.066	8.265	60.7	22.4	0.889
SR3	-	5.082	8.244	60.9	22.5	0.793
SR4	-	5.081	8.235	63.2	23.4	1.219
SR5	-	5.079	8.240	63.4	23.4	1.158

Table II. Crude Oil, Produced Water, and Caustic Solution Properties

Crude Oil	
Density (g/cc)	0.8842
Viscosity (Pa.s) @ 22.2°C	0.0292
API Gravity (deg)	28.4
Acid number (mg KOH/g crude)	1.75
Produced Water	
Ion	Concentration (g/kg)
Calcium	0.0180
Magnesium	0.0130
Sodium	1.8640
Sulfate	0.6160
Chloride	2.4890
Total Dissolved Solids (TDS)	5.0000
Viscosity (Pa.s) @ 22.2°C	0.00125
Density (g/cc)	1.0019
pH	7.2
Caustic Solution	
Concentration (g/kg)	5.0
Density (g/cc)	1.0070
Viscosity (Pa.s) @ 22.2°C	0.0014
pH	13.3

reservoir brine prepared in the laboratory. The produced water (PW) also served as the solvent for the polymer and alkaline solutions. Moreover, it was the driving fluid during the waterflooding and postflushing phases of each coreflood.

Since the alkaline agent/connate water interactions were not the major concern of this study, it was decided to eliminate the magnesium hydroxide and calcium hydroxide precipitation. Therefore, the PW was softened by filtering through a resin column saturated with sodium (Na) and chloride (Cl) ions. Thus, the divalent cations (Mg^{++} and Ca^{++}) in the PW were exchanged with the monovalent cations (Na^{+}). Note that the term "produced water (PW)" will be used to refer to the softened produced water in the remainder of this paper.

Alkaline Solution

The interfacial tensions (IFT) between the crude oil and eleven caustic, NaOH, solution samples of different concentrations were measured by a Spinning Drop Tensiometer. The lowest IFT recorded was 0.08-mN/m at a caustic concentration of 0.5-g/kg.

Nevertheless, the NaOH concentration used in the alkaline injection solution and the alkaline/polymer blend solution was chosen as 5.0-g/kg, ten times higher than the NaOH concentration yielding the lowest IFT. Such a high caustic concentration could, therefore, provide excess caustic to be consumed via ion exchange and rock dissolution reactions during the propagation of caustic pulse through the core. Additionally, this would compensate for the further dilution of the caustic-containing slug with the connate water within the core.

It should be remembered that the term "caustic solution" will stand for the alkaline injection solution having a NaOH concentration of 5.0-g/kg throughout the rest of this paper.

Chase Polymer and Alkaline/Polymer Solutions

The polymer used in this study was American Cyanamid Company's 30-percent hydrolyzed Cyanatrol 950S anionic polyacrylamide polymer with a molecular weight of $14x10^{6}$. The hydrodynamic radius of gyration, R_g, of the polymer molecule, which was found to be 0.32-μm, was calculated by using the following Fox-Flory equation[27]

$$R_g = \{10^{14}\ MW\ ([\eta]/\Phi)\}^{1/3} \tag{1}$$

where MW is the polymer molecular weight, Φ is a universal constant given by[26]

$$\Phi = (10/3)\ \pi\ N\ \xi^3 \tag{2}$$

where N is Avagadro's number, and ξ is a universal constant equal to 0.875 for high molecular weight macromolecules, according to the Kirkwood-Riseman theory[28]. The intrinsic viscosity, $[\eta]$, in Eqn. (1) can be determined from

$$[\eta] = \{(\mathrm{Ln}\ \eta_r)/c\}_{c=0} \tag{3}$$

$$\eta_r = \eta'/\eta_s \tag{4}$$

where η_r, η', and η_s are the relative viscosity, zero shear viscosity and solvent viscosity, respectively. Reference 9 contains the details and the application of the above procedure.

A series of capillary pressure measurements was conducted by mercury injection into 1-inch Berea sandstone core plugs, which had similar physical properties to those used in the coreflooding experiments. Then, the capillary pressure data obtained were utilized to determine the average pore radius, which was found to be 3.5-μm[9].

As a rule-of-thumb, the magnitude of R_g should be less than one third the average pore radius in order to eliminate core plugging problems during polymer injection. Since the average pore radius and R_g data obey the rule-of-thumb, Cyanatrol 950S was confirmed to be an appropriate polymer to be injected into the cores without confronting pore plugging and injectivity loss problems.

Table III. Injection Sequences of the Corefloods

Core	WF (PV)	NaOH (PV)	30HPA/NaOH (PV)	30HPA (PV)	PF (PV)
SL1	1.0	0.4	-	-	1.5
SL2	1.0	-	-	0.5	1.5
SL3	1.0	-	-	1.0	1.5
SL4	1.0	-	0.2	0.8	1.5
SL5	1.0	-	0.4	0.6	1.5
LL1	1.0	-	0.2	0.8	1.5
LL2	1.0	-	0.4	0.6	1.5
LR1	1.0	-	0.2	0.8	1.5
LR2	1.0	-	0.4	0.6	1.5
SR1	1.0	0.4	-	-	1.5
SR2	1.0	-	-	0.5	1.5
SR3	1.0	-	-	1.0	1.5
SR4	1.0	-	0.2	0.8	1.5
SR5	1.0	-	0.4	0.6	1.5

WF : waterflooding

NaOH : caustic injection

30HPA : polymer injection

30HPA/NaOH : injection of the blend of polymer and caustic

PF : second and final waterflooding (postflush)

PV : slug size, pore volume

The polymer solution was prepared by diluting the 30,000-ppm oil-external polyacrylamide stock solution to a concentration which yielded a viscosity in the range of 0.001 to 0.004-Pa.s higher than the crude oil viscosity. The polymer was inverted to be water external either in PW or in caustic solution by shearing in a blender for 20 seconds. The polymer solution inverted in PW was the "chase polymer solution", which is designated as 30HPA, and the polymer solution inverted in caustic solution was the "blend solution of alkaline and polymer," which is designated as 30HPA/NaOH.

Both 30HPA and 30HPA/NaOH solutions were filtered through a 200-mesh screen to remove "fish eyes," which are small oil-external polymer gels which did not invert to be water external. Very few fish eyes were encountered, indicating that the preparation technique for polymer containing solutions was successful.

EXPERIMENTAL PROCEDURE

Initially, the cores were evacuated under a vacuum pressure of 20-millitorr for at least twelve hours, and then saturated with PW. The initial oil saturation (S_{oi}) and irreducible water saturation ($S_{w\text{-}irr}$) conditions in each core was obtained by injecting at least 2.0 PV of crude oil. The crude oil injection was stopped when there was no more water production from the core and the differential pressure across the core was stabilized.

The establishment of S_{oi} and $S_{w\text{-}irr}$ conditions was followed by the waterflooding, EOR, and postflush phases, respectively. The sequences of the types and sizes of the injected slugs during each coreflood are given in Table III, and the schematic of the coreflooding unit is illustrated in Figure 3. As seen in Table III, each core couple underwent the same injection sequences, so that the results could be compared.

The frontal velocities in the linear cores and the median frontal velocities in the radial cores were set to approximately 0.3048-m/day in order to simulate typical frontal velocities encountered in actual field floods. The Reynolds number (Re), which should be less than 1.0 for Darcy flow in the core, was calculated to be much less than 0.01 for each coreflood, even when the largest grain diameter of 0.1-cm (determined by an SEM study) was considered. All experiments were conducted at room temperature of 22°C.

RESULTS AND DISCUSSION

Oil Recovery

Oil recoveries obtained from waterflooding (WF) and subsequent EOR phases of each coreflood are tabulated in Table IV. Also, the predicted ultimate waterflood oil recoveries which could be reached after infinite water throughput are included in Table IV. Moreover, Figure 4 displays the comparison of the oil recovery efficiencies of different processes applied in small radial cores.

It is obvious from the results given in Table IV and Figure 4 that the second version of the PAAF process is superior to both the AF and PF processes in enhancing the oil recovery. In addition, the larger the size of the 30HPA/NaOH slug the higher the oil recovery. It was also

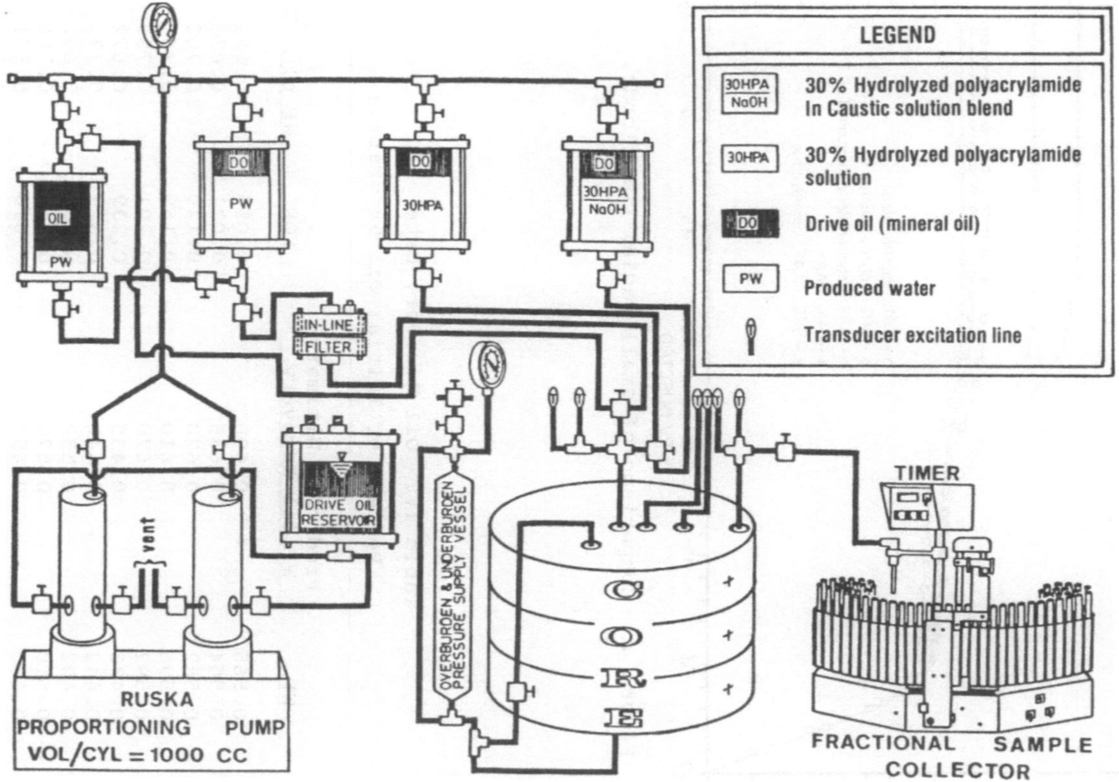

Figure 3. Schematic diagram of coreflooding system.

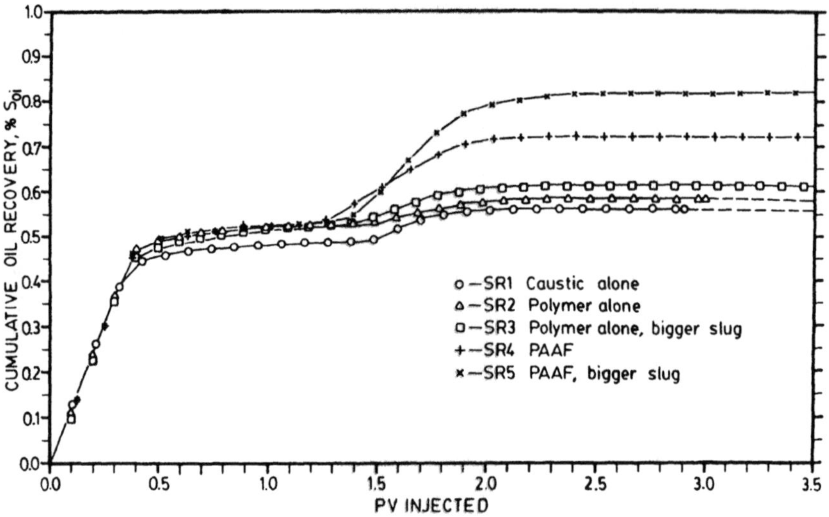

Figure 4. Comparison of small radial corefloods.

Table IV. Oil Recovery Data
Percent of Initial Oil Saturation

Core	WF	Predicted Ultimate WF Oil Recovery	EOR	WF plus EOR
SL1	0.565	0.600	0.056	0.621
SL2	0.564	0.635	0.071	0.635
SL3	0.582	0.620	0.120	0.702
SL4	0.596	0.610	0.169	0.765
SL5	0.592	0.670	0.201	0.793
LL1	0.433	0.435	0.239	0.672
LL2	0.442	0.442	0.329	0.771
LR1	0.492	0.550	0.200	0.692
LR2	0.510	0.576	0.216	0.726
SR1	0.485	0.492	0.076	0.561
SR2	0.522	0.530	0.062	0.584
SR3	0.519	0.535	0.093	0.612
SR4	0.523	0.532	0.198	0.721
SR5	0.530	0.540	0.290	0.820

determined that the larger size of the 30HPA/NaOH slug decreases the relative overall caustic consumption.

Polymer Viscosity Reduction

During the preparation of alkaline/polymer blend solution (30HPA/NaOH) samples it was observed that the viscosity of the sample was decreasing continuously in time. Therefore, the addition of more polymer into the caustic solvent became a necessity in order to set the viscosity of the 30HPA/NaOH sample to a value slightly higher than the crude oil viscosity. While adding more polymer, a time of 10-hours had to elapse for the viscosity reduction to level off and become relatively constant.

Consequently, it was realized that the viscosity behavior of the 30-percent hydrolyzed Cyanamid 950S polyacrylamide polymer in caustic environment was following a trend very similar to that followed by the PHPA polymers in Figure 1. Such an outcome was nothing but the confirmation of Szabo's results analyzed previously in this paper.

Observation of a New Effect

Each core was inspected after the completion of displacement experiments. Unexpectedly, a visible residual oil saturation (ROS) ring (adjacent to the zone which was totally cleaned out of oil near the injection sand face) was observed in both radial and linear cores which underwent the PAAF process. On the contrary, such an effect did not occur in the cores flooded with either alkaline or polymer solution alone. Figures 5.a and 5.b illustrate this newly recognized effect in radial and linear cores, respectively.

It appears that the thickness of the ROS ring is a direct function of the 30HPA/NaOH slug size. In other words, the larger the size of the 30HPA/NaOH slug the thicker the ROS ring. For comparison purposes, as demonstrated in Table III, the size of the 30HPA/NaOH slug was kept constant as either 0.2 or 0.4 PV, relative to the pore volume in each core. Thus, the effect of core size on the size of the ROS ring was not investigated for scaling to actual field floods.

Also, the authors did not attempt to determine the residual oil saturation in the ring. However, the color intensity inspection on the ROS ring indicated that the amount of residual oil in the ring was lower than the initial saturation in this portion of the core.

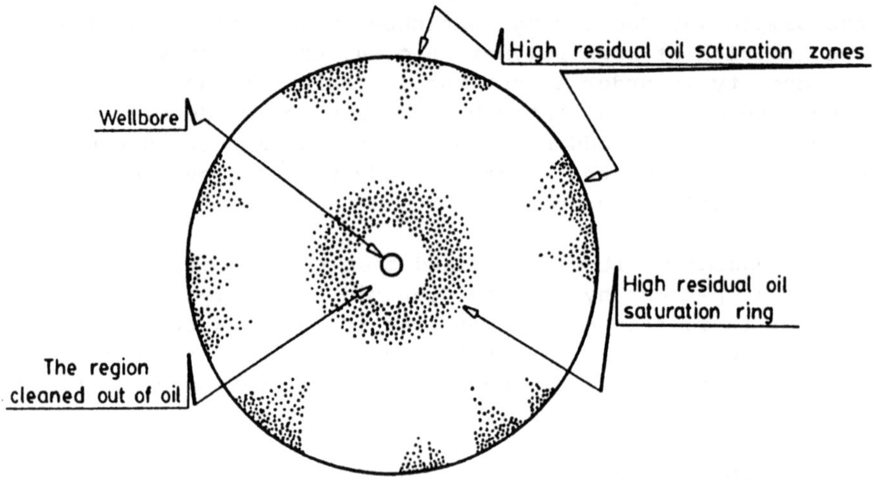

Figure 5.a. Schematic illustration of the residual oil saturation ring in a radial core.

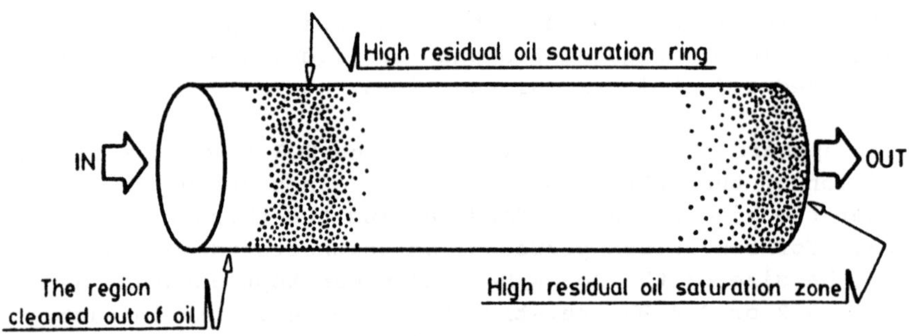

Figure 5.b. Schematic illustration of the residual oil saturation ring in a linear core.

Nevertheless, the residual oil saturation in the ring appeared to be higher than that in the other swept portions of the core.

A developed hypothesis, incorporating the stability of PHPA polymers in a caustic environment, is explained below for the phenomena causing the formation of the ROS ring during the application of the PAAF process.

Consider a relatively homogeneous sandstone core which was flooded with water initially. The core has a water-flooding residual oil saturation ($S_{or\text{-}wf}$), which is less than the initial oil saturation (S_{oi}), and an average water saturation ($S_{w\text{-}wf}$), which is higher than the irreducible water saturation ($S_{w\text{-}irr}$). Also, consider that this core is going to be flooded with a 30HPA/NaOH slug to be chased by a fresh 30HPA slug.

The injection of the 30HPA/NaOH slug cleans the oil completely out of a small portion of the rock adjacent to the injection sand face. The pore volume (PV) in this cleaned portion is very similar to the volume of the 30HPA/NaOH slug (see Figure 6.a).

One of the reasons for the thorough cleaning in this small portion of the core is that the 30HPA/NaOH slug has not yet been affected by the microscopic heterogeneities of the sandstone. The other reason is that the alkaline agent and crude oil can react very effectively, due to the high level of alkalinity and thorough contact. Thus, the saponification, IFT reduction, temporary wettability alteration, and/or emulsification mechanisms progress almost ideally. Additionally, a simultaneous and a very efficient volumetric sweep is provided by the polymer, which has not yet lost a significant number of its molecules by adsorption and entrapment.

However, the polymer retention mechanism, adsorption and mechanical entrapment of polymer molecules, immediately but slowly start to denude the 30HPA/NaOH of polymer. In conjunction with this, the leading edge of the 30HPA/NaOH slug starts becoming unstable, because it feels the dispersing effect of the microscopic heterogeneities of the porous medium. Therefore, almost ideal sweep and displacement efficiencies cannot be maintained, and the 30HPA/NaOH slug, which is slowly being stripped of polymer, advances through the core while leaving some residual oil behind its leading edge, as seen in Figure 6.b. Thus, the formation of the ROS ring starts at this point.

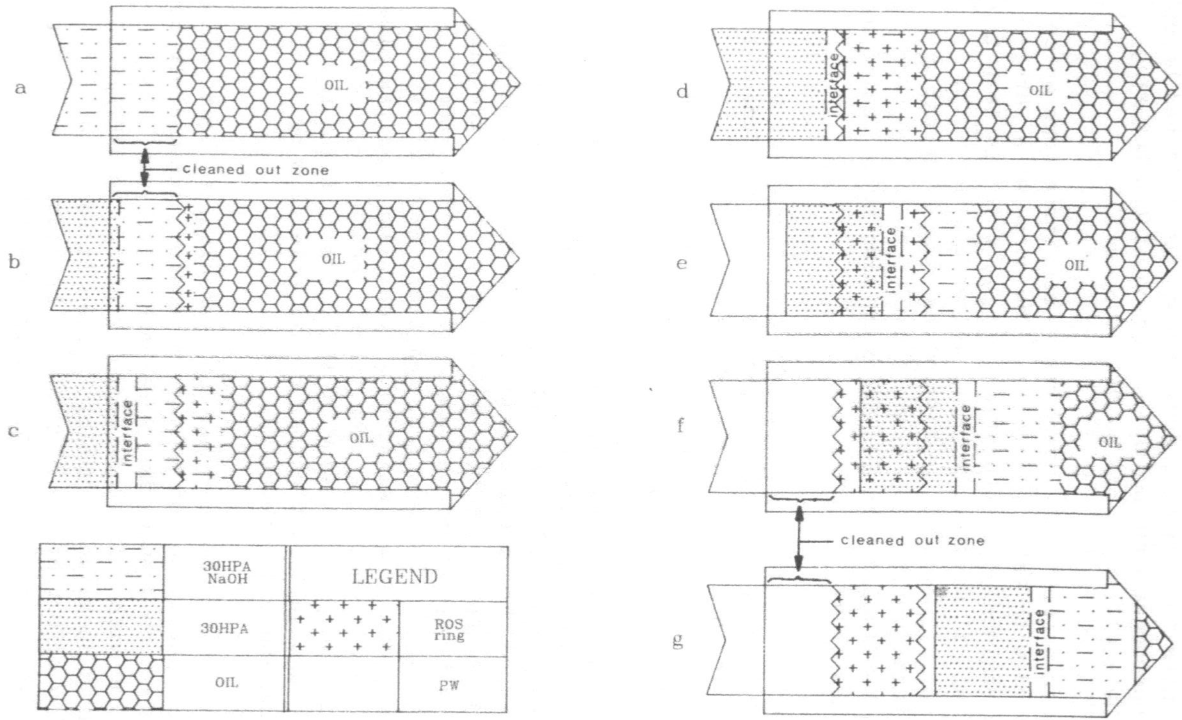

Figure 6. Schematic illustration of the residual oil saturation ring formation in a linear core.

The displacement front of the 30HPA/NaOH slug, which has passed the cleaned out portion (see Figure 6.c), becomes more unstable as it feels the effects of physical dispersion and polymer retention mechanisms more strongly. The instability is reflected in two interrelated modes: (1) the decrease in sweep and displacement efficiencies at the leading edge (displacement front) of the slug due to (2) the variation in polymer concentration distribution throughout the slug. Hence, the instability in the 30HPA/NaOH slug promotes the formation of the ROS ring by leaving more residual oil behind the leading edge.

The variation in polymer concentration distribution in the slug can be described as the change from an initial isotropic distribution to an anisotropic distribution. The anisotropic distribution displays a gradual increase in polymer concentration from zero at the leading edge of the 30HPA/NaOH slug to its maximum value at the trailing edge. It should be noted that this maximum value of the polymer concentration may be less than its original value.

The occurance of such a gradual distribution of polymer concentration in the slug can be explained as follows: All the polymer molecules in the 30HPA/NaOH slug exhibit a tightly wound coil structure, due to the pseudo salt effect of caustic. The polymer molecular coils at the leading edge of the slug are retained continuously and effectively to form a mono or multilayer coating on the pore walls, depending on the geometry and ion exchange potential of the adsorption and entrapment sites. Thus, the polymer concentration at the leading edge would be zero upon retention at any point in time during the formation of the ROS ring.

On the other hand, the trailing edge of the slug has a high concentration of polymer molecular coils, because it passes through the previously coated pores without loosing its polymer coils to the retention mechanism. Note that the local flow velocity becomes high at the high polymer retention sites due to the reduced pore radius. Such an increase in local velocity causes the polymer molecular coils at the trailing edge to accelerate and eventually catch up with the leading edge of the slug.

In other words, the trailing edge feeds the leading edge through the middle part of the slug, and thus the concentration of the polymer molecular coils increases from zero at the leading edge to its maximum at the trailing edge. However, the polymer concentration in the 30HPA/NaOH slug decreases continuously due to the polymer retention at the leading edge, while the ROS ring is being formed.

In this situation, a relatively inefficient mobility control is pronounced at the leading edge, and thus the oil displacement is provided mainly by the caustic phase. Therefore, the formation of the ROS ring continues as seen in Figures 6.b through 6.d.

In the meantime the injection of the 30HPA/NaOH slug is completed, and the chasing fresh polymer slug (30HPA) starts entering the cleaned out portion of the core (Figure 6.b). The 30HPA slug exhibits a higher viscosity and better mobility control, and a rapid interreaction starts between its leading edge and the trailing edge of the 30HPA/NaOH slug while forming a transitional interface (Figure 6.c). The viscosity in the transitional interface starts decreasing due to the pseudo salt effect of caustic.

During the propagation of the transitional interface through the cleaned out portion of the core, a small number of the polymer molecules may be adsorbed and entrapped to complete the saturation with polymer in this region. When saturation is completed, the pore walls in the cleaned out portion are coated and the small pore throats are plugged with polymer molecules. Therefore, the chasing 30HPA slug behind the transitional interface passes through this cleaned out portion without losing its polymer molecules and the mobility control capability (Figures 6.c and 6.d).

The chasing 30HPA slug, therefore, starts feeding the 30HPA/NaOH slug with somewhat coiled up polymer molecules through the transitional interface. Eventually, the distribution of polymer concentration within the 30HPA/NaOH slug starts regaining its initial feature and good mobility control capability. At this moment, the formation of the ROS ring becomes completed, as illustrated in Figures 6.d and 6.e.

Although the mobility control in the 30HPA/NaOH slug is significantly decreased during the ROS ring formation, some amount of waterflooding residual oil in this ring was still mobilized. Hence, the final residual oil saturation in the ring is somewhat lower than the S_{or-wf}, which was the oil saturation in the ring region at the end of the water-flooding phase.

Once the formation of the ROS ring is completed, the volumetric sweep and displacement efficiencies, mobility control, and oil recovery increases significantly in the rest of the core farther downstream (Figures 6.e through 6.g). The polymer requirement for the polymer retention in the rest of the core is provided mainly by the 30HPA slug through the transitional interface.

Detection of the Transitional Interface

The Reciprocal Relative Mobility (RRM) concept[29] was utilized to study the mobility behavior in cores which underwent the PF and PAAF processes. The following RRM equations were used for each of the core geometries:

Linear core: $$RRM = (6.803*10^{-5})\ [kA(\Delta p)/QL] \qquad (5)$$

Radial Core: $$RRM = (4.274*10^{-4})\ [kh(\Delta p)/Q\ Ln(r_e/r_w)] \qquad (6)$$

Figures 7 through 10 illustrate some of the PV Injected versus RRM plots generated. The propagation of a lower viscosity transitional interface through the cores which underwent the PAAF process is clearly visible and shown by the arrows in Figures 7 and 8. On the other hand, such a behavior cannot be detected in Figures 9 and 10, which display the viscosity behavior in the cores flooded with polymer solution (30HPA) alone.

Significance of the ROS Ring

When the second version of the PAAF process is applied in an actual oil reservoir, it is most likely that the ROS ring will form some distance away from the injection well. Since the size of the ROS ring will be very large in a field scale, the amount of oil left in the ROS ring may become economically significant.

Even though the application of the first version of the PAAF process may seem to be more appropriate in order to eliminate the possible formation of the ROS ring, it would not be the right decision. First of all, the second version of the PAAF process can recover at least the same amount of oil as the first version of the PAAF process could recover, with less caustic consumption. Second, it is believed that conducting a dilute polymer preflush before the start of the second version of the PAAF process may eliminate the

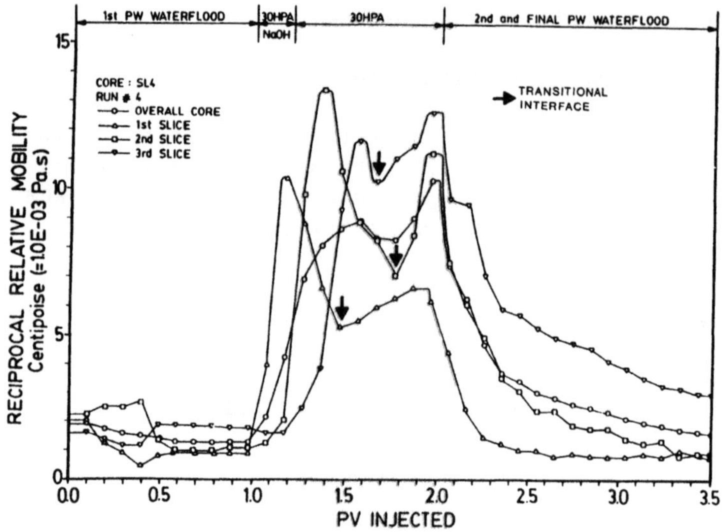

Figure 7. The history of mobility behavior in Core SL4.

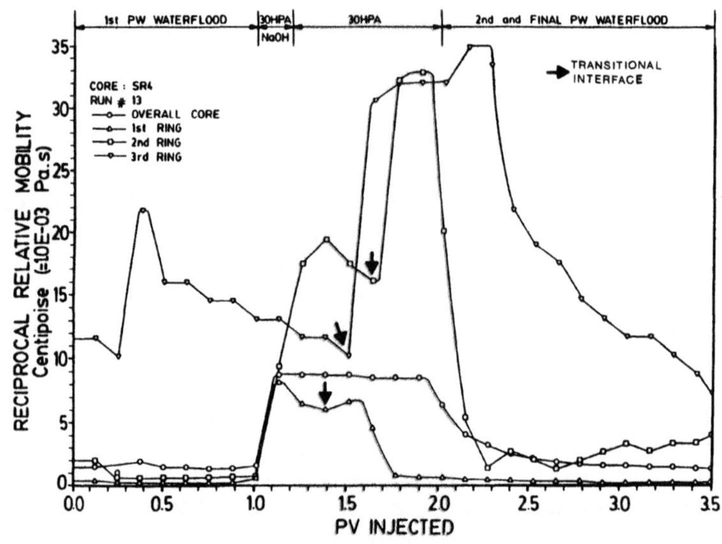

Figure 8. The history of mobility behavior in Core SR4.

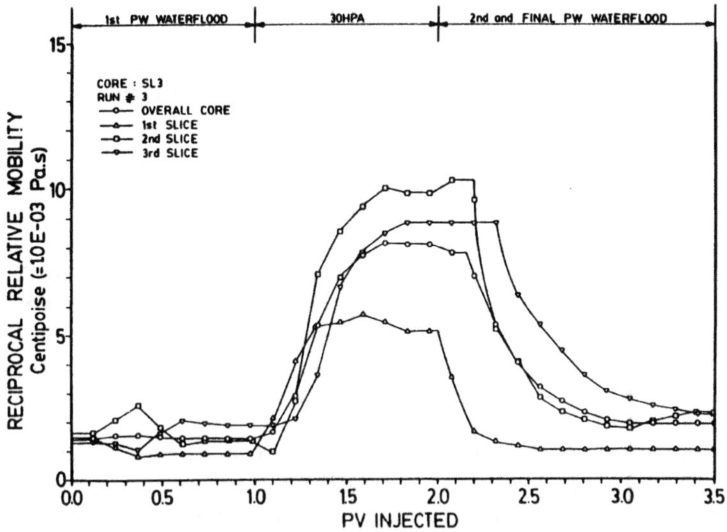

Figure 9. The history of mobility behavior in Core SL3.

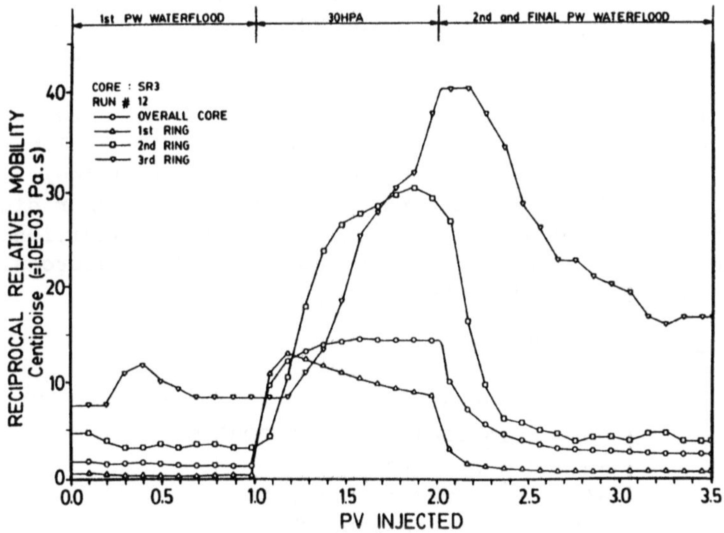

Figure 10. The history of mobility behavior in Core SR3.

formation of the ROS ring by providing a polymer precoating on the pore walls.

CONCLUSIONS

1. The second version of polymer augmented alkaline flooding (PAAF) is superior to both alkaline flooding and polymer flooding alone in enhancing oil recovery.

2. The viscosity of the partially hydrolyzed polyacrylamide polymers decreases continuously in caustic media, and levels off after a long time. The time required for the viscosity to become constant, which could be in the order of several hours or several days, depends on the initial hydrolyzation degree and the nature of the polymer used, and the concentration of salts and caustic in the solution.

3. A residual oil saturation (ROS) ring forms a little away from the injection sand face, when the second version of the PAAF process is applied in cores of both radial and linear flow geometries. The residual oil saturation in this ring seems to be lower than what it was initially in that region in the core.

4. It is believed that conducting a dilute polymer preflush could eliminate the formation of the ROS ring by providing a polymer precoating on the pore walls.

5. During the application of the second version the PAAF process a transitional interface of relatively low viscosity is formed between the leading alkaline/ polymer blend slug and the chasing fresh polymer slug. The existance and the propagation of the transitional interface can be detected by applying the reciprocal relative mobility (RRM) concept.

NOMENCLATURE and UNITS

A - cross-sectional area of a linear core, cm^2
AF - conventional alkaline flooding
EOR - enhanced oil recovery
h - hight of a radial core, cm
IFT - interfacial tension, mN/m
k - absolute permeability, md
L - length of a linear core, cm
LL - long-linear core plug
Ln - natural logarithm function

LR	-	large-radial core disk
MW	-	molecular weight
N	-	Avagadro's number
NaOH	-	caustic soda
NH_2	-	amide group of a polyacrylamide molecule
p	-	pressure, psi
PAAF	-	polymer augmented alkaline flooding
PF	-	polymer flooding
PHPA	-	partially hydrolyzed polyacrylamide
PV	-	pore volume, cm^3
PW	-	produced water
Q	-	volumetric flow rate, cm^3/s
r_e	-	outer radius of a radial core, cm
R_g	-	hydrodynamic radius of gyration of a polymer molecule, μm
ROS	-	residual oil saturation
RRM	-	reciprocal relative mobility, cP
r_w	-	radius of the wellbore in a radial core, cm
SEM	-	scanning electrone microscope
SL	-	short-linear core plug
S_{oi}	-	initial oil saturation, fraction
S_{or-wf}	-	waterflood residual oil saturation, fraction
SR	-	small-radial core disk
S_{w-irr}	-	irreducible water saturation, fraction
S_{w-wf}	-	average water saturation after the waterflood, fraction
WF	-	waterflooding
30HPA	-	30-percent hydrolyzed polyacrylamide polymer solution
30HPA/NaOH	-	blend solution of 30-percent hydrolyzed polyacrylamide polymer and caustic

Greek Letters

Δ	-	difference
π	-	ratio of the circumference to the diameter of a circle
Φ	-	a universal constant, dimensionless
ζ	-	a universal constant, dimensionless
$[\eta]$	-	intrinsic viscosity of a polymer solution, dl/g
η'	-	zero-shear viscosity of a polymer solution, cSt
η_r	-	relative viscosity of a polymer solution, dimensionless
η_s	-	solvent viscosity of a polymer solution, cSt

UNIT CONVERSION FACTORS

atm	=	(1.315 789 E-06) x millitorr
centipoise	=	(1.0 E+03) x Pa.s
dyne/cm	=	(1.0 E+00) x mN/m
°F	=	(°C x 9/5) + 32
ft/day	=	(3.048 E-01) x m/day
millidarcy	=	(1.013 274 E+03), μm^2
Pa	=	(9.869 23 E+0.6) x atm
psi	=	(1.450 326 E+03) x kPa

REFERENCES

1. Farmanian, P.A., Davis, N., Kwan, J.T., Weinbrandt, R.M., and Yen, T.F.: "Participation of Selective Native Petroleum Fractions in Lowering Interfacial Tensions of Aqueous Alkaline," Chemistry of Oil Recovery, edited by Johansen, R.T., and Berg, R.L., ACS Symposium Series No. 91, American Chemical Society, Washington D.C., 1979, pp: 103-114.

2. Squires, F., U.S. Patent No. 1,238,355 (1917).

3. Mungan, N.: "Enhanced Oil Recovery Using Water as a Driving Fluid, Part 4 - Fundamentals of Alkaline Flooding," World Oil, Vol. 192, No. 6, June 1981, pp: 209-220.

4. Mungan, N.: "Enhanced Oil Recovery Using Water as a Driving Fluid, Part 5 - Alkaline Flooding Field Applications", World Oil, Vol. 193, No. 1, July 1981, pp: 181-190.

5. Mayer, E.H., Berg, R.L., Carmichael, J.D., and Weinbrandt, R.M.: "Alkaline Injection for Enhanced Oil Recovery - A Status Report," Jour. of Pet. Tech., Vol. 35, No. 1, January 1983, pp: 209-221.

6. Novosad, Z., and Novosad, J.: "Determination of Alkalinity Losses Resulting from Hydrogen Ion Exchange in Alkaline Flooding," Soc. of Pet. Eng. Jour., Vol. 24, No. 1, February 1984, pp: 49-52.

7. Lieu, V.T., Miller, S.G., and Staphanos, S.J.: "Long Term Consumption of Caustic and Silicate Solutions by Petroleum Reservoir Sands," Soluble Silicates, edited by J.S. Falcone, Jr., ACS Symposium Series No. 194, 1982, pp. 227-250.

8. Garrett, B.T., Krumrine, P.H., and Van Kirk, C.W.: "Alkaline Flooding - Design of a Low Cost, Portable Facility," paper SPE 11847 presented at the SPE-AIME Rocky Mountain Reg. Meet., Salt Lake City, Utah, May 1983.

9. Mihcakan, I.M.: "The Effects of Polymer Augmented Alkaline Flooding and Core Geometry on Ultimate Oil Recovery," M. Sc. Thesis, T-2902, Colorado School of Mines, Golden, Colorado, May 1985.

10. Mungan, N.: "Enhanced Oil Recovery Using Water as a Driving Fluid, Part 8 - Application of Polymer Solutions to Improve Waterflooding", World Oil, Vol. 194, No. 2, February 1, 1982, pp: 95-106.

11. Sloat, B., and Zlomke, D.: "The Isenhour Unit - A Unique Polymer-Augmented Alkaline Flood," paper SPE/DOE 10719 presented at the SPE/DOE 3rd Joint Symp. on Enhanced Oil Recovery, Tulsa, Oklahoma, April 1982.

12. Burk, J.H.: "Comparison of Sodium Carbonate, Sodium Hydroxide and Sodium Orthosilicate for EOR," paper SPE 12039 presented at the SPE-AIME 58th Annu. Tech. Conf., San Francisco, California, October 1983.

13. Ball, J.T., and Pitts, M.J.: "Simulation of Reservoir Permeability Heterogeneities with Laboratory Corefloods," paper SPE 11790 presented at the SPE-AIME International Symp. on Oilfield and Geothermal Chemistry, Denver, Colorado, June 1983.

14. Ball, J.T., and Pitts, M.J.: "Effect of Varying Polyacrylamide Molecular Weight on Tertiary Oil Recovery from Porous Media of Varying Permeability," paper SPE 12650 presented at the SPE/DOE 4th Symp. on Enhanced Oil Recovery, Tulsa, Oklahoma, April 1982.

15. Sandiford, B.B.: "Flow of Polymers Through Porous Media in Relation to Oil Displacement," Improved Oil Recovery by Surfactant and Polymer Flooding, edited by Shah, D.O., and Schecter, R.S., Academic Press, Inc., San Francisco, California, 1977, pp: 491-495.

16. Jennings, R.R., Rogers, J.H., and West, T.J.: "Factors Influencing Mobility Control by Polymer Solutions," Jour. of Pet. Tech., Vol. 23, March 1971, pp: 391-401.

17. Mungan, N., Smith, F.W., and Thompson, J.L.: "Some Aspects of Polymer Floods," Jour. of Pet. Tech., Vol. 18, No. 10, September 1966, pp: 1143-1150.

18. Mungan, N.: "Shear Viscosities of Ionic Polyacrylamide Solutions," Soc. of Pet. Eng. Jour., Vol. 12, No. 6, December 1972, pp: 469-473.

19. Martin, F.D., and Sherwood, N.S.: "The Effect of Hydrolysis of Polyacrylamide on Solution Viscosity, Polymer Retention and Flow Resistance Properties," paper SPE 5339 prepared for the SPE-AIME Rocky Mountain Reg. Meet., Denver, Colorado, April 1975.

20. Maerker, J.M.: "Shear Degradation of Partially Hydrolyzed Polyacrylamide Solutions," Soc. of Pet. Eng. Jour., Vol. 15, No. 4, August 1975, pp: 311-322.

21. Ferrer, J.: "Some Mechanistic Features of Flow of Polymers Through Porous Media," paper SPE 4029 prepared for the SPE-AIME 47th Annu. Tech. Conf., San Antonio, Texas, October 1972.

22. Szabo, M.T.: "An Evaluation of Water-Soluble Polymers for Secondary Oil Recovery - Part 1," Jour. of Pet. Tech., Vol. 31, No. 5, May 1979, pp: 553-560.

23. Thomas, C.P.: "The Mechanism of Reduction of Water Mobility by Polymers in Glass Capillary Arrays," paper SPE 5556 presented at the SPE-AIME 50th Annu. Tech. Conf., Dallas, Texas, September 1975.

24. Hirasaki, G.J., and Pope, G.A.: "Analysis of Factors Influencing Mobility and Adsorption in the Flow of Polymer Solution Through Porous Media," Soc. of Pet. Eng. Jour., Vol. 14, No. 4, August 1974, pp: 337-346.

25. Szabo, M.T.: "Some Aspects of Polymer Retention in Porous Media Using a C^{14}-Tagged Hydrolyzed Polyacrylamide," Soc. of Pet. Eng. Jour., Vol. 15, No. 4, August 1975, pp: 323-337.

26. Chauveteau, G.: "Molecular Interpretation of Several Different Properties of Flow of Coiled Polymer Solutions Through Porous Media in Oil Recovery Conditions," paper SPE 10060 presented at the SPE-AIME 56th Annu. Tech. Conf., San Antonio, Texas, October 1981.

27. Billmeter, F.W., Jr.: "Textbook of Polymer Science, 2nd Ed.", Wiley-Interscience, a Division of John Wiley and Sons, Inc., New York City, 1971, pp: 27-30, 84-90.

28. Tanford, C.H.: "Physical Chemistry of Macromolecules," John Wiley and Sons, Inc., New York City, 1965, pp. 390-400.

29. Argabright, P.A., and Rhudy, J.S.: "Partially Hydrolyzed Polyacrylamides with Superior Flooding and Injection Properties," paper SPE 11208 presented at the SPE-AIME 57th Annu. Tech. Conf., New Orleans, Louisiana, September 1982.

AUTHOR INDEX

SUBJECT INDEX

MIX
Papier aus verantwortungsvollen Quellen
Paper from responsible sources
FSC® C105338

If you have any concerns about our products,
you can contact us on
ProductSafety@springernature.com

In case Publisher is established outside the EU,
the EU authorized representative is:
Springer Nature Customer Service Center GmbH
Europaplatz 3, 69115 Heidelberg, Germany

Printed by Libri Plureos GmbH
in Hamburg, Germany